Sensors
Principles and Applications

Sensors
Principles and Applications

Peter Hauptmann
Translated by Tim Pownall

Carl Hanser Verlag

Prentice Hall

First published in German in 1991 by Carl Hanser Verlag, Munich, under the title *Sensoren : Prinzipien und Anwendungen* by Peter Hauptmann.

This edition first published in English in 1993 by Carl Hanser Verlag, Munich and Prentice Hall.

Prentice Hall International (UK) Ltd
Campus 400, Maylands Avenue
Hemel Hempstead
Hertfordshire, HP2 7EZ
A division of
Simon & Schuster International Group

© Carl Hanser Verlag 1991

All rights reserved. No part of this publication may be reproduced, stored in a retrieval system, or transmitted, in any form, or by any means, electronic, mechanical, photocopying, recording or otherwise, without the prior permission, in writing, from the publisher.
For permission within the United States of America contact Prentice Hall Inc., Englewood Cliffs, NJ 07632

Typeset in 10/12 pt English Times
by Mathematical Composition Setters Ltd., Salisbury, UK

Printed in Great Britain by Redwood Books,
Trowbridge, Wiltshire

Library of Congress Cataloging-in-Publication Data

Hauptmann, Peter.
  [Sensoren. English]
  Sensors : principles and applications / Peter Hauptmann ; translated by Tim Pownall.
    p.   cm.
  Includes bibliographical references and index.
  ISBN 0-13-805789-3
  1. Detectors. I. Title.
TA165.H34 1993   92-42438
681′ .2--dc20   CIP

British Library Cataloguing in Publication Data

A catalogue record for this book is available from the British Library

ISBN 0-13-805-789-3 (pbk)

---

1 2 3 4 5 97 96 95 94 93

Contents

Preface ix

List of abbreviations xi

List of symbols xiii

1 Introduction 1

2 Definition of the term 'sensor' 4

3 Techniques in sensor production 8

4 Silicon sensors 9
    4.1 Properties of silicon and their effects on sensors 9
    4.2 Production stages in silicon technology 11
    4.3 Micromechanical processes 13
    4.4 Temperature sensors 16
        4.4.1 Resistance temperature sensors 16
        4.4.2 Interface temperature sensors 18
        4.4.3 Other silicon temperature sensors and applications 21
    4.5 Pressure sensors 22
        4.5.1 The piezoresistive effect 23
        4.5.2 Piezoresistive pressure sensors 24
        4.5.3 Capacitive pressure sensors 30
        4.5.4 New pressure sensor principles 31
    4.6 Optical sensors 34
        4.6.1 Photoresistors 36
        4.6.2 Photodiodes and phototransistors 37
        4.6.3 Photodiode arrays 42
        4.6.4 Charge coupled devices 44
        4.6.5 Other semiconductor materials for optical sensors 47

vi Contents

    4.7 Magnetic field sensors — 48
        4.7.1 Galvanomagnetic effects — 49
        4.7.2 Hall generators and magnetoresistors — 53
        4.7.3 FET-Hall sensors, magnetic diodes and transistors — 56
        4.7.4 Possible applications of magnetic field sensors — 60
    4.8 Micromechanical sensors — 63
        4.8.1 Acceleration/vibration sensors — 63
        4.8.2 Microbridge sensors — 65

5 Sensors in thin-film technology — 68

    5.1 Deposition techniques — 69
    5.2 Selected thin-film sensors — 69

6 Thick-film sensors — 75

    6.1 Production stages — 75
    6.2 Examples — 76

7 Fiber optic sensors — 79

    7.1 The structure of fibers — 79
    7.2 The classification of fiber optic sensors — 81
    7.3 Applications — 83
        7.3.1 Multimode sensors — 83
        7.3.2 Monomode sensors — 93
            7.3.2.1 Monomode sensors with phase modulation — 93
            7.3.2.2 Monomode sensors with polarization modulation — 102
        7.3.3 Distributed fiber optic sensors — 105
        7.3.4 Other selected fiber optic sensor techniques — 109
    7.4 New fibers — 112

8 Chemical sensors — 115

    8.1 Detection principles and chemical sensor requirements — 116
    8.2 Design types — 119
        8.2.1 Conductivity sensors — 120
        8.2.2 Structured semiconductor sensors — 124
        8.2.3 Electrochemical sensors — 126
        8.2.4 Solid electrolyte sensors — 133
        8.2.5 Chemically sensitive FETs (CHEMFETs) — 138
        8.2.6 Specific designs — 142
            8.2.6.1 Optrodes — 142
            8.2.6.2 Biosensors — 145
            8.2.6.3 Humidity sensors — 150

9	**Sensors based on 'classical' measuring elements**	154
	9.1 Inductive sensors	154
	9.2 Capacitive sensors	157
	9.3 Ultrasound sensors	159
	9.4 Other principles	167
10	**New sensor materials**	168
	10.1 Piezoelectric polymers	169
	10.1.1 Fabrication and properties	169
	10.1.2 Applications	172
	10.2 Amorphous metals	174
	10.2.1 Magnetic field sensors	175
	10.2.2 Magnetoelastic sensors	178
	10.2.3 Other sensor principles	180
	10.3 The Wiegand sensor	182
11	**Resonance sensors**	185
	11.1 Quartz resonance sensors	186
	11.2 Surface acoustic wave sensors	192
	11.3 Resonance sensors for density, filling level and throughflow	197
12	**Prospects for the future**	201
	References	203
	Index	213

Preface

The stimulus for this book came from the lack of a text which covered the most important work on sensor development and applications for graduate students, engineers in industry and research scientists.

Nowadays sensors play a dominant role in many spheres of our daily life. They are included in consumer products, cars, medical devices and aeroplanes. Process monitoring and control could not be realized without the use of a great variety of sensors. Air pollution is detected by special gas sensors. Military devices, e.g. rockets, are equipped with sensors. Many other examples could be listed.

The sensor converts information about the environment such as temperature, pressure, force, etc., into an electrical signal. Sometimes the first piece of information is an optically coded signal which is converted in a second step into an electrical signal. This appears in the case of fiber optic sensors.

With the development of microelectronics in the 1970s new sensors became more and more interesting. In particular, they made use of the properties of silicon. By using microelectronic technologies cheap, miniaturized sensors were produced. New sensor materials were discovered and known or new principles were introduced for practical aims. The integration of sensor and signal conditioning electronics created exciting opportunities for a great number of applications. Thus a new era in the sensor field began in the 1980s. From a scientific point of view this was demonstrated by the rapid growth of published papers, new journals, and conferences on sensors. In industry many small and medium-sized sensor development and production companies started up in the 1980s.

Nowadays the miniaturization of sensors is the aim of many research laboratories and companies. As a part of microsystem technology sensors also have a major role to play in the future. With the development of new materials, sophisticated technologies and new ideas for sensor principles, the sensor field will become more significant in years to come. Many producers in Japan, Europe, and the USA forecast growth rates above 10 per cent beyond the year 2000.

Despite the importance of sensors, books about them are rare. There are two reasons for this: first, the rapid pace of development; and second, the interdisiciplinary nature of the subject. Electronic engineers, physicists and chemists have to work together. This book attempts to satisfy all the demands of these different disciplines. It describes the most important types of sensor – semiconductor sensors; fiber optic sensors; chemical/biological sensors; and resonant sensors. The importance of new sensor materials and principles is shown by diverse examples. A brief summary of technologies such as silicon technology, thin- and thick-film techniques and micromachining is also given. Physical descriptions of the different sensor types are given. Examples are given of their technical applications. The opportunities for their use as well as their limitations are described. Interesting laboratory developments are indicated which can play a dominant role in the future.

This book is based on the author's extensive experience in sensor research (e.g. ultrasonic sensors, chemical sensors, new sensor materials), lectures on sensors at the Technical University of Magdeburg and industrial applications of his own sensor developments. It is addressed to engineers, natural scientists in research and industry, and graduate students with an interest in this fascinating subject.

List of abbreviations

ASIC	application-specific integrated circuit
ASIS	application-specific integrated sensor
BCCD	buried charge coupled device
CCD	charge coupled device
CIM	computer integrated manufacturing
CMOS	complementary metal oxide semiconductor
CVD	chemical vapour deposition
EMI	electromagnetic interference
ENFET	enzymatic field-effect transistor
FET	field-effect transistor
FFT	fast Fourier transform
FOS	fiber optic sensor
FPR	Fabry–Pérot resonator
HF	high frequency
IC	integrated circuit
IOC	integrated optical chips
ISE	ion-sensitive electrode
ISFET	ion-sensitive field-effect transistor
ITO	indium tin oxide
LED	light emitting diode
LPCVD	low pressure chemical vapour deposition
LVDT	linearly variable differential transformer
LWS	Lamb wave sensor
MFS	magnetic field sensors
MIS	metal–isolator semiconductor
MOS	metal oxide semiconductor
MT	magnetotransistor
NDT	non-destructive testing
NMOS	N-channel metal oxide semiconductor
NMR	nuclear magnetic resonance
NTC	negative temperature coefficient
OPV	operational amplifier

OTDR	optical time domain reflectometry
PCVD	pressure chemical vapour deposition
PMMA	polymethylmetacrylate
POTDR	polarization optical time domain reflectometry
PRESSFET	pressure sensitive field-effect transistor
PTC	positive temperature coefficient
PTFE	poly(tetrafluoroethene)
PVDF	poly vinylidene difluoride
PZT	piezoceramic transformer
QMB	quartz microbalance
SMD	surface mounted devices
SOI	silicon on isolator
SOS	silicon on sapphire
SQUID	superconducting quantum interference device
TCR	temperature coefficient of resistance
VLSI	very large scale integration
WSG	wire strain gauge

List of symbols

A	area	I_C	collector current
A	amplitude	I_F	forward current
$A(\lambda, U)$	spectral sensitivity	I_S	saturation current
a	activity	I_{CE}	collector–emitter current
B	magnetic induction	I_{CB}	collector–base current
C	electrical capacitance	k	k factor
c_o	speed of light *in vacuo*	k	Boltzmann constant
c	speed of sound	k	anisotropy constant
c	concentration	l	length
c_{11}	coefficient of elasticity	m	mass
D	thickness	M	moment of force
d	distance	M	magnetization
d_{ij}	dielectric coefficient	N	number of turns
E	energy	NA	numeric aperture
E	elastic modulus	n	refractive index
e	elementary charge	n	reaction value
F	force	n_i	intrinsic conductivity
F	Faraday constant	p	pressure, partial pressure
F_L	Lorentz force	p	sound pressure
F_C	Coriolis force	Q	partial function
$F(\lambda)$	energy spectrum	Q	mechanical quality
f	frequency	q	electric charge
G	free enthalpy	R	electrical resistance
G	conductance	R	reflection factor
G	geometry factor	R	radius
G	shear modulus	R_H	Hall constant
g_{ij}	piezoelectric constant	r	radius
H	magnetic field strength	s	recombination speed
h	height	T	transmission factor
h	Planck constant	T	absolute temperature
I	electric current	t	time
I	intensity	t_p	delay time

List of symbols

U	electrical voltage	θ	angle
U_H	Hall voltage	θ_H	Hall angle
U_F	forward voltage	λ	wavelength
U_{BE}	base–emitter voltage	λ_s	saturation magnetostriction
v	carrier velocity		
v	velocity (flow)	μ	relative permeability
w	width	μ_n	(carrier) mobility
W	energy	μ_o	induction constant
W_g	energy difference	ν	frequency
x, y, z	cartesian coordinates	π_L, π_T	piezoresistive coefficient
Z	acoustic impedance	ρ	density
α	sound absorption	ρ	specific resistivity
α	angle of polarization	σ	voltage, mechanical stress
α	angle of rotation	τ	decay, relaxation time
α'	temperature coefficient	τ	duration
β	compressibility	Φ	phase
δ	thickness of diffusion layer	Φ	ionization energy
		Φ	magnetic flux
ε_0	absolute dielectric constant	φ	phase
ε_r	relative dielectric constant	φ	angle
η_s	anomalous viscosity	φ	potential
η_v	kinematic viscosity	ω	angular frequency
θ'	degree of coverage	Ω	angular velocity

1 Introduction

Sensors represent a qualitatively new stage in the more effective exploitation of all the possibilities which have been made available by microelectronics, in particular in the field of information processing. Sensors form the interface between the electronic control system on the one hand, and the environment, the process, procedure or machine on the other. In the past, sensor development has been unable to keep pace with the rate of development in the microelectronics industry. Indeed, at the end of the 1970s and the beginning of the 1980s sensor development was internationally considered to be between three and five years behind that of the microelectronics sector. The fact that microelectronic components were frequently much cheaper to manufacture than the measuring elements (sensors) they required was a serious obstacle to increasing and diversifying the application of information processing microelectronics in a wide range of processes and procedures. This discrepancy between modern microelectronics and classical measuring technology could only be eliminated through the development of modern sensor technology. For this reason, sensors are now regarded as one of the key elements for the continued and accelerated development of microelectronics.

Intensive research and development work in the various fields of sensor technology commenced on an international basis. The result of this activity is that today's sensor market enjoys one of the highest annual growth rates (between 10 and 20 per cent). As sensors form the basis for gaining all the information required in connection with process statuses and environments (in the widest sense of the term), they open the door to completely new possibilities for the automation of a range of processes in industry, the household, the environment, medical applications and other sectors. For example, the fully automated, integrated factory of the future can only be realized with the help of sensors.

Although sensors, in conjunction with information processing microelectronics, represent an important step forward, it is only the first such step. At this stage they make use of a range of available microelectronic components, for example in the form of processors, memories, analog–digital converters or amplifiers, to prepare the output signal. At the same time, the sensor has to provide an electronic output which is easy to process, ideally in the form of a digital, bus-compatible signal. There is also the need for miniaturization.

2 Introduction

The second step is the sensor–microelectronics–actuator connection. The information gained by the sensor concerning the status or progress of a process passes through an electronic signal processing stage to an actuator (classically a controller) which feeds back to the process. The sensor–microelectronics–actuator chain only works if all its links are compatible. This leads to the specification of further important criteria, especially as far as the sensor is concerned. Despite widespread acknowledgement of the significance of the sensor as one of the key elements in the automation process, it is difficult to obtain comprehensive, comparative information about the state of sensor technology and developments in the field. There are a number of reasons for this:

1. Sensors are available for the measurement of more than 100 physical dimensions. If we also take the measurement of chemical species into account, this number increases to many hundreds.
2. It is possible to classify approximately 2000 basic types of sensor [1]. Between 60 000 and 100 0000 sensors for on-line process measurements are commercially available in the western world.
3. According to the INSPEC database, more than 10 000 sensor-related publications appear each year.
4. Generally speaking, the development of new sensors or sensor technologies takes an estimated 5–15 years, and is a very cost-intensive process.

It might be concluded that it is a hopeless task to attempt to write a book about sensors unless, as in [2,3], one restricts oneself to a single specific sector of this wide-ranging subject. However, it is precisely for this reason that it seems so necessary to provide both developers and users of sensors with an overview of what already exists and the developments that can be expected in the future. Thus although a sensor developer working in a specific field might consider his own interests to be dealt with too quickly or in insufficient detail, he should be stimulated to push his thoughts beyond the boundaries of his own discipline.

The sensor field has a highly interdisciplinary character. From the developer through to the user, a large number of professions are represented. This fact has its advantages; for example, it allows for the complex application of one or more sensor principles. However, there are also disadvantages, such as the level of resistance to the introduction of new, unknown ideas.

In the following I shall attempt to address the entire field, presenting both the physical premises and the technologically relevant applications of the most important principles of sensors. Since the multiplicity of sensors makes it impossible to take a single technology or a particular measurand as the connecting thread of this work, it has been necessary to select another procedure which will satisfy the criteria of breadth and depth of presentation.

The first part of the book deals with sensors which are produced using microelectronic technology. This is followed by a discussion of fiber optic and chemical sensors, that is to say, sensors in which the focus is placed on the relevant principles on the one hand, and on the dimension to be measured on the other. Next come sensors based on so-called classical measurement principles and those based on new materials which open up new possibilities. Because of their significance, resonance sensors are dealt with separately. A short discussion of the most important development trends concludes the book.

This layout means that a particular measurand, temperature or displacement for example, is dealt with in a number of different chapters. This would appear to be a disadvantage. However, it is advantageous in that it makes possible a clear definition of the production-technology-dependent limits of application of different sensors.

The current book extends the range of works published in recent years by authors who have had the same objectives and intentions but have enjoyed varying levels of success [2–9]. Given the frenetic rate of growth in the sensor field, the current book is at least differentiated by the fact that is is completely up to date.

2 Definition of the term 'sensor'

Today the word 'sensor' in no way lags behind concepts such as microprocessor, transputer, the various designations for memory and other electronic components as one of the buzzwords of technological innovation. However, it still lacks an exact definition such as has long existed for terms like 'probe', 'gauge', 'pick-up' or 'transducer'. It is therefore not surprising that publications dealing with sensors frequently commence with a definition. There have been many attempts to restrict this plethora of terms [10–16]. Apart from the word 'sensor', we also find sensor element, sensor system, intelligent or smart sensor, sensor technology, and so on. What is it that is hidden behind the word 'sensor'? It is a specialist term derived from the Latin word *sensorium*, meaning 'sensory capability', or *sensus*, meaning 'sense'. Given this origin of the concept of sensors, it seemed obvious to emphasize the analogy between technical sensors and the human sense organs. Figure 2.1 presents this analogy. However, the idea of a sensor goes beyond this analogy to become an all-embracing synonym for the sensing, conversion and recording of measured values.

Before we embark on our discussion of sensors we shall need to define the term. The following definition will be used:

> A sensor converts (changes in) the physical dimension which is to be measured into (changes in) an electrical dimension which can be processed or transmitted electronically.

The physical dimensions can be classified in accordance with the diagram in Figure 2.2. Table 2.1 gives examples of the physical dimensions which sensors can measure.

The sensor can be subdivided into a sensing element which, for example, registers pressure as the deflection of a semiconductor membrane, or a change in refractive index as the fall in light intensity in a fiber optic; as well as a converter or transduction element which converts the deflection of the semiconductor membrane, in which resistors are diffused in bridge form, into an electrical voltage, or translates the change in light intensity into an electrical signal by means of an optoelectronic conversion process [2]. A sensor may also consist of the transduction element alone (for example, piezoelectronic sensors, photosensors). This definition

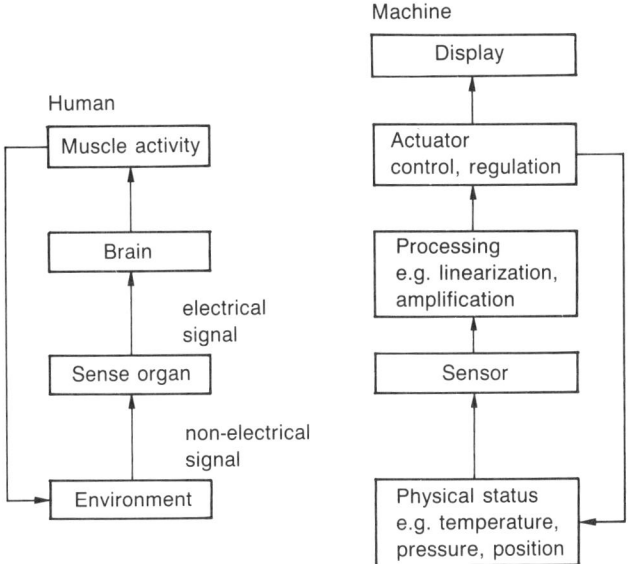

Figure 2.1 There is an obvious comparison between the sense organs which allow humans to apprehend their environment and the sensors produced by technology. Although there are vast differences, there are also often remarkable similarities. What is common to both is that the sensor or sense organ itself often has little to tell us. It is the intelligent system that creates the information.

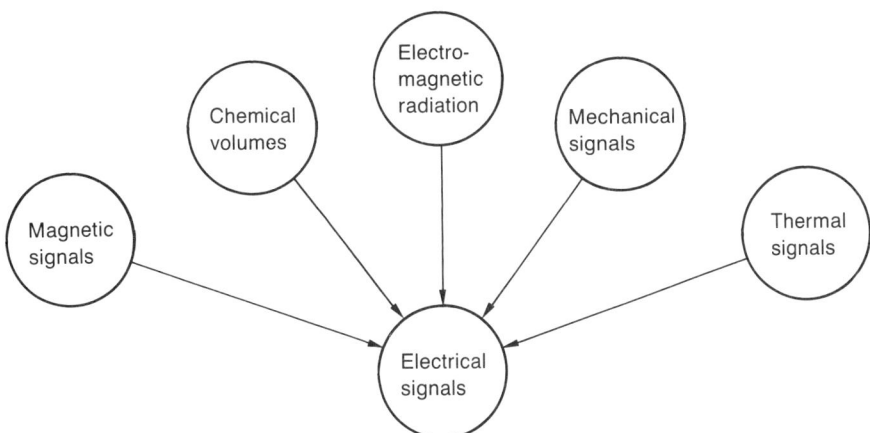

Figure 2.2 Physical dimensions which are converted into an electrical signal by a sensor.

6 Definition

Table 2.1 Sensor-measurable quantities

MECHANICAL DIMENSIONS OF SOLIDS
separation acceleration elasticity density thickness torque speed of revolution pressure diameter shape filling level speed weight power length height hardness mass orientation throughflow tension distance angle etc.

MECHANICAL DIMENSIONS OF LIQUIDS AND GASES
density pressure viscosity volume velocity of flow throughflow etc.

THERMAL DIMENSIONS
temperature heat flow thermal radiation etc.

OPTICAL RADIATION
intensity wavelength polarization reflection colour etc.

ACOUSTIC DIMENSIONS
sound pressure velocity of propagation absorption intensity sound frequency etc.

NUCLEAR RADIATION
radiant energy degree of ionization radiant flux etc.

CHEMICAL SIGNALS
pH value concentration molecule or ion type particle form and size reaction speed humidity etc.

MAGNETIC AND ELECTRICAL SIGNALS
inductance capacitance resistance frequency phase current voltage permittivity magnetic field strength etc.

OTHER IMPORTANT DIMENSIONS
quantity pulse duration time etc.

---

of the sensor places no restrictions on its size or shape. This is intentional since I shall later discuss both miniaturized and more bulky sensor units.

The real difference between classical measuring instruments and sensors can be seen in the next stage, namely in the preparation and processing of the electrical signal which is always the objective of a sensor. The result is that there are fundamental differences in the requirements that classical measuring instruments and sensors have to meet [4]. It also justifies the use of the term 'sensor', although there will, of course, always be a greater or lesser degree of overlap.

The signal preparation stage includes, for example, amplification, filtering, analog–digital conversion or simple correction circuits. The electronics responsible for signal preparation may be either integrated with the sensor or spatially separate from it. In the first case we speak of *integrated sensors*. Otherwise the sensor and the signal preparation electronics together are frequently known as a *sensor system* [2, 14]. This distinction is illustrated in Figure 2.3. However, the term 'sensor system' may also refer to the connection of multiple sensors of the same or different type. In the case of some sensors (for example, tactile or optical), the term *array* is frequently used to describe this latter case. If the following electronic stage makes use of processors which allow the implementation of correction algorithms, diagnostics, tests or the selective polling of different sensors, then we speak of

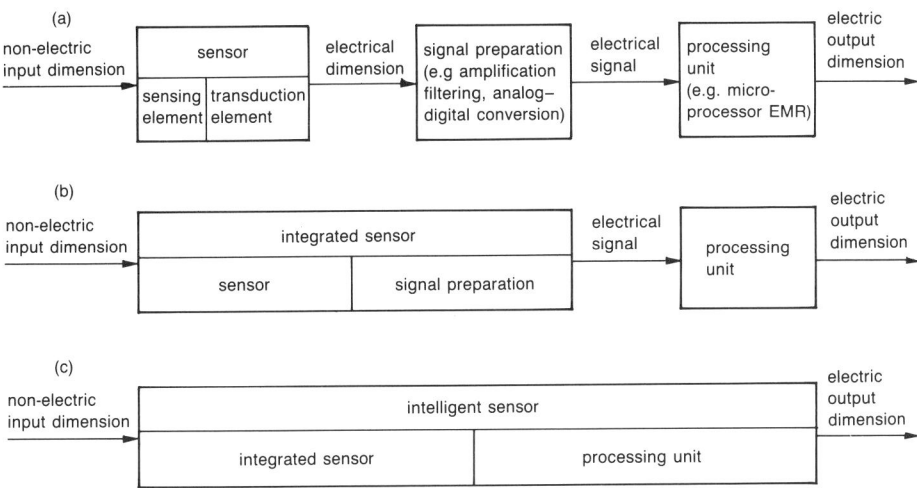

Figure 2.3 Possible types of sensor system: (a) sensor system with discrete construction; (b) sensor system with integrated sensor; (c) intelligent sensor system or intelligent sensor.

intelligent sensors. Even if the idea of intelligence is greatly exaggerated in comparison with the human faculty, it nevertheless expresses a relatively high level of signal processing. This capability requires a sufficiently intelligent software, which is usually referred to as *smart*.

The term *sensor technology* refers to the totality of the sensor or sensor system with its preparation and processing of signals in both the associated hardware and software. Sensors must fulfil general requirements if they are to interest users. A good sensor should possess the following characteristics [17]:

- adequate sensitivity;
- high degree of accuracy and good reproducibility;
- high degree of linearity;
- good dynamic range;
- insensitivity to interference and environmental influences;
- high degree of stability and reliability;
- long life expectancy and problem-free replacement.

The combination of sensor and sensor electronics leads to specific requirements, such as:

- noise-free output signal;
- bus-compatible output signal;
- low power requirements.

If the sensor and the electronics are to be integrated then further requirements must be met, such as miniaturizability and cost compatibility with microelectronics.

3 Techniques in sensor production

Current sensor technology is still based on a relatively large number of non-miniaturized sensors. This can be seen by analyzing the geometrical dimensions of sensors for the measurands distance, power, acceleration, velocity, throughflow, pressure, and so on [18]. For many sensors, these dimensions exceed 10 cm. Frequently the dimensions of the sensor are not determined by the sensor itself but by that of its housing. However, even in such cases the sensors themselves are several centimetres in size. Such sensors, which can sometimes be very expensive, will remain important in the future, for example in the fields of process measurement, production technology and robotics. However, parallel to this it is possible to observe another development which has been triggered by advances in microelectronics. Microelectronic technology has stimulated the development of sensors which are miniaturizable and suitable for mass production environments. This most certainly does not mean that sensor technology will develop at the same pace as microelectronics. The aim of miniaturization offers a range of benefits. For example, the effect of a miniaturized sensor on the measured parameters is weak. This means that the sensor introduces a lesser degree of interference and therefore achieves a higher degree of measurement precision. The inertia of the sensor is reduced and it consumes less electrical power than classical sensors. The following microelectronics technologies are employed especially frequently:

- silicon technology;
- thin-film technology;
- thick-film/hybrid technology;
- other semiconductor technologies (III-V and II-VI semiconductors).

Other processes are also employed in sensor production, such as foil or sinter technologies, fiber optic technology, precision mechanics, optical laser technology, microwave technology and biological technologies. In addition, technologies involving polymers, metal alloys or piezoelectric materials also play a role in sensor production.

Because silicon and other semiconductors are used so extensively in microelectronics, I shall first examine those sensors which are based on semiconductor technologies. In each case, I shall present a brief outline of the technology before turning my attention to the sensors themselves.

4 Silicon sensors

The preferred strategy in the development of new silicon-based sensors is to exploit the techniques and processes that are already established within the silicon-based integrated circuit (IC) industry and in this way to benefit from the experience and achievements of this industrial sector. Since equally great advances have been made in the growth of silicon crystals, the range of possibilities for the production of silicon sensors is growing steadily. Nowadays it is the use of monocrystalline silicon that is the focus of attention for sensor applications. However, both polycrystalline silicon (poly-Si) and amorphous silicon (a-Si) are of considerable interest and show great promise for the future. Thin layers of these are deposited on to substrates. We can expect the use of such materials to increase as soon as the deposition processes can be better controlled.

The use of the various effects of silicon in sensors requires a wide-ranging modification of the parameters of the monocrystalline bulk material or the silicon layers. These parameters may include the concentration of impurities, the lattice purity, the grain size of poly- or a-Si as well as the doping level. The effect of these parameters on the various silicon types is presented in detail in [2].

4.1 Properties of silicon and their effects on sensors

Silicon is a suitable material for sensor technologies if it manifests sufficient physical and chemical effects of an acceptable strength which can be used in uncomplicated structures across a wide range of temperatures. Table 4.1 presents the most important effects and their applications for sensor technology. Frequently such considerations impose exacting requirements on the properties and structure of the material. Further requirements may result from the subsequent development step to intelligent sensors.

The use of silicon has a number of implications for sensors. Firstly, the physical properties of silicon can be used directly to measure the desired measurand, as indicated in Table 4.1. However, the range of possibilities is limited. Beyond this,

Table 4.1 Silicon effects used in sensors

Physical measurand	Effect	Application
Radiation	photoresistive effect	photoresistor
	photointerface effect	photodiode, phototransistor, CCD array and matrix MIS and Schottky diode
	ionization effect	nuclear radiation sensor
	photocapacitive effect	photocapacitance
Mechanical quantity	piezoresistive effect, piezojunction and piezotunnel effect	piezoresistive power and pressure sensors, piezoelectric diode and transistor
Thermal quantity	thermal resistance	resistance temperature sensors
	thermojunction effect	temperature sensors (diode, transistor)
	thermoelectric effect	thermopile
	pyroelectric effect	pyroelectric sensor
Magnetic signals	magnetoresistive effect	magnetoresistive sensors
	Hall effect	Hall generator
	magnetic interface effect	magnetic diode and transistor
Chemical signals	charge-sensitive field effect	ISFET

silicon can, for example, be extremely useful when used as the substrate for thin-film sensors, even when information processing electronics are integrated. Under certain circumstances, modification of the silicon modules in the semiconductor electronics makes it possible to develop important types of sensor. Optical sensors, such as photodiodes, or chemical sensors, such as ISFETs, are examples of this. The latest development, which emerged at the beginning of the 1980s, is the coupling of microelectronic technologies with techniques developed specially for sensor production, such as anisotropic wet etching or anodic glass on silicon bonding. In this way the excellent mechanical properties of monocrystalline silicon can be used to develop innovative sensors. This technology, which is known as *micromechanics*, leads to the production of mechanical or mechanical/electronic silicon components with dimensions similar to their electronic counterparts, being a few tens of a micrometer in size. Monocrystalline silicon is particularly well suited to this technology on account of its excellent mechanical properties (Table 4.2). Monocrystalline silicon does not creep. However, its brittleness can be a disadvantage. Like diamond, the crystal can be fractured on various planes. Further restrictions are caused by the band-gap of 1.1 eV which means that intrinsic conduction occurs at high temperatures. p-n junctions and all the sensors which are

Table 4.2 Properties of silicon in comparison with other materials

	Tensile strength 10^9 N/m²	Young modulus 10^{11} N/cm²	Density g/cm³	Thermal conductivity W/cm °C
TiC^x	20	4.97	4.9	3.3
Al₂O₃	15.4	5.3	4	0.5
Si^x	7	1.9	2.3	1.57
Steel	4.2	2.1	7.9	0.97
Al	0.17	0.7	2.7	2.36

^{x}-monocrystal

based on them then cease functioning. Finally the dopant becomes diffused. The result is that many sensors based on monocrystalline silicon are limited to applications in which the temperature does not rise above 120–150°C.

Amorphous and polycrystalline silicon layers on insulating substrates make a broader range of modifications to characteristic data possible and enhance the potential for integration. As a result, in comparison with monocrystalline silicon, it is possible to increase the measuring range, to higher temperatures for example, and to compensate for interference, for example by reducing the offset. The sensitivity of the monocrystalline silicon material, which is primarily determined by the resistivity ρ, the mobility μ and the lifetime τ of the minority carriers, depends to a large extent on the selection of the starting material. In contrast, the sensitivity of poly-Si and a-Si can be greatly modified during layer deposition and processing. The use of poly-Si and a-Si opens up interesting new areas of application for sensors.

4.2 Production stages in silicon technology

The fabrication of silicon sensors is largely based on processes employed in modern semiconductor technology which have been developed for the production of microelectronic components. For example, lithography used in conjunction with coating and doping techniques makes it possible to determine the structure of materials right down to the micrometre or sub-micrometre range. Silicon planar technology not only dominates the production of integrated circuits, but is also the decisive element in the production of many silicon sensors. This leads to benefits such as:

- low-cost manufacture of sensors in large quantities;
- miniaturization of the sensor and the monolithic integration of the sensor and the electronics;
- creation of multisensors (multiple sensors on a single chip).

12 Silicon sensors

However, in the case of silicon sensors particular attention should be paid to the following peculiarities [19]:

- the use of large chips or, in some cases, whole wafers (e.g., solar cells, position-sensitive photoelectric sensors);
- the possibility of three-dimensional structuring in which special techniques for deep and anisotropic etching and special etch stop layers are used to create the three-dimensional form of miniaturized silicon components (so-called 'micromechanical processes' which are of considerable significance for today's new sensor developments);
- the use of very thin discs or very thin parts (pressure or acceleration sensors);
- the depositing of thin sensor layers on a silicon substrate which extends the limited sensor properties of silicon while simultaneously exploiting its advantages such as the ability to integrate electronics, etc.;
- the packaging methods must meet sensor-specific requirements which are very different from the methods used for microelectronic components (finding suitable encapsulation methods is one of the greatest problems of silicon sensor manufacture).

In the following I shall present only a brief outline of the production steps involved in silicon planar technology. A detailed presentation can be found in [2]. The characteristics of micromechanics will be discussed separately.

The base material both for the fabrication of silicon components and for micromechanics is provided by wafers of monocrystalline silicon. Typically these have a diameter of between 3 and 6 inches with a thickness of 300–600 μm. The most commonly used wafers have a (100) orientation. They can be n- and p-doped.

The key processing stages in silicon technology can be subdivided as follows:

- generation of insulation layers;
- doping of semiconductor layers or epitaxial growth of new layers;
- structuring of the layers and the substrate;
- metallization;
- bonding;
- encapsulation and protection.

SiO_2, Si_3N_4 and Al_2O_3 can be used for the insulating layers. The silicon base material is usually doped by the diffusion of impurity atoms or by ion implantation. Because of its many advantages, implantation has now widely replaced the diffusion method. It is used to produce p-n junctions, layers of differing conductivity and etch stop layers.

An epitaxial process makes it possible to deposit monocrystalline silicon with the required doping on a single-crystal substrate such as silicon or sapphire. These epitaxial layers improve the characteristic data of the components.

The generation of suitable structures on a silicon wafer is usually performed using a lithographic process. While the microelectronics sector frequently makes use of electron-beam and X-ray lithography, sensors and micromechanical components are still usually produced using conventional photolithography. The binding of the various layers, the provision of ohmic contacts and the connection of bonding contacts are performed through the application of aluminium metallization. The individual elements are then equipped with leads (bonded) and encapsulated.

Protection and encapsulation are important for any sensor and give rise to conflicting requirements. Recently, increasing attention has sensibly been paid to developing suitable bonding, protection and encapsulation methods.

When manufacturing sensors, it may also be necessary to apply additional sensitive layers. These may be ferroelectric, ion-sensitive, moisture-sensitive or piezoactive layers. Polycrystalline and amorphous silicon may also be used. Their properties are determined by the method used to deposit the layers, the doping and activation of the charge carriers as well as the recrystallization of the poly-Si [2, 10, 11, 22, 23, 24]. The LPCVD and PCVD processes are now commonly used to deposit layers of poly-Si. Amorphous layers are generally deposited using hydrogen in CVD processes, and it is from this that the designation 'a-Si:H' comes. However, such layers can also be produced by means of sputtering techniques.

4.3 Micromechanical processes

The term 'micromechanics', with its obvious similarities to the term 'microelectronics', is used to describe a completely new discipline. Its objective is the construction of complex microsystems consisting of largely integrated sensors, a logical signal processing stage and actuators. In this connection, the term 'micromechanics' is used to refer to the fabrication of mechanical structures whose geometrical size is, at least in one dimension, so small that it is no longer sensible to use the methods of fine mechanics. Depending on the boundary conditions imposed by the desired function or by the properties of the material, this limit may be located anywhere between the millimetre and the sub-micrometre range (see Figure 4.1). In contrast to microelectronics, micromechanics is concerned with the production of three-dimensional structures.

Crystalline silicon is the prevalent material used in micromechanics. Apart from the advantage conferred by the direct monolithic integration of mechanical properties and electronics on a single chip, the so-called 'anisotropic etching process' represents another considerable factor in its favour. This method makes use of wet etching for silicon with a speed of etching that is highly dependent on the doping (selectivity) and the crystal orientation (anisotropy) [26]. Orientation-dependent etching can be performed, for example, using the ethylenediamines KOH, NaOH, H_2O_2. While only attacking the densest atomic packings in the (111) orientation very mildly, they etch surfaces with orientations of (100) or (110) very rapidly. The ratio of the etching rate at orientations of (100) and (111) can greatly exceed 100:1 (Figure 4.2). Surfaces with an orientation of (111) therefore present

14 Silicon sensors

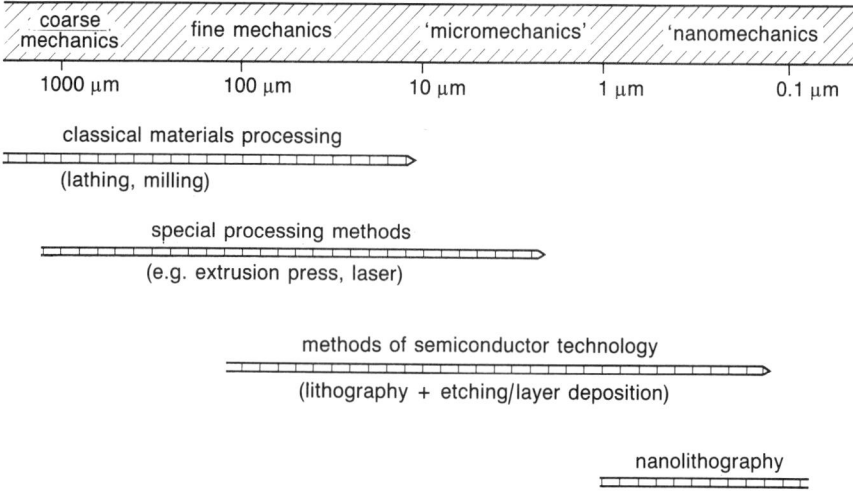

Figure 4.1 Illustration of the term 'micromechanics'.

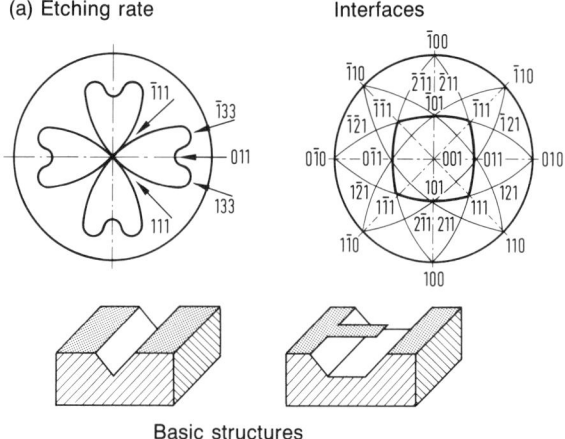

Figure 4.2 Orientation-dependent etching speeds for silicon wafers with (100) orientation (after [26]).

a barrier to etching. They are vertically aligned and can be used to define specific structures. They form an angle of 54.7° with the surface of a wafer with (100) orientation. This means that in (100) material they produce a self-delimiting 'V' profile. In (110) material they produce vertical sides. Using the appropriate masking it is possible to produce very exact pit-shaped channels or an inverted right-angle pyramid structure. Only at convex faces is an underetching of the structures possible. In fact, given suitable masking it should be possible to produce a wide variety of structures (Figure 4.3). As the etching speeds for anisotropic etching are considerably lower than for isotropic etching, it is not possible to use waxes or photoresists. The mask structure must be transferred from a photoresist to an SiO$_2$ layer.

Boron-doped layers are used as the etch stop layer. A reduction in the speed of etching by a factor of up to 1000 is observed at boron concentrations of 10^{20} atoms/cm^3. Layers with a high level of boron doping can be produced using an epitaxial process, for example. Such layers are useful for the production of membranes, cantilever beams and helices. Doping to the required level makes it possible to terminate the etching process at the required boundary. More rarely, chemical methods can be used to terminate etching [28, 29].

Using the methods described above it is possible, starting with a wafer with a thickness of 300–500 μm, to produce a pressure-sensitive membrane with a thickness of 10–20 μm with precisely controlled lateral dimensions and a variation in thickness of less than 1 μm [30].

A second method which has recently become of considerable interest for micromechanics is the selective etching of silicon on isolator (SOI) structures. In this

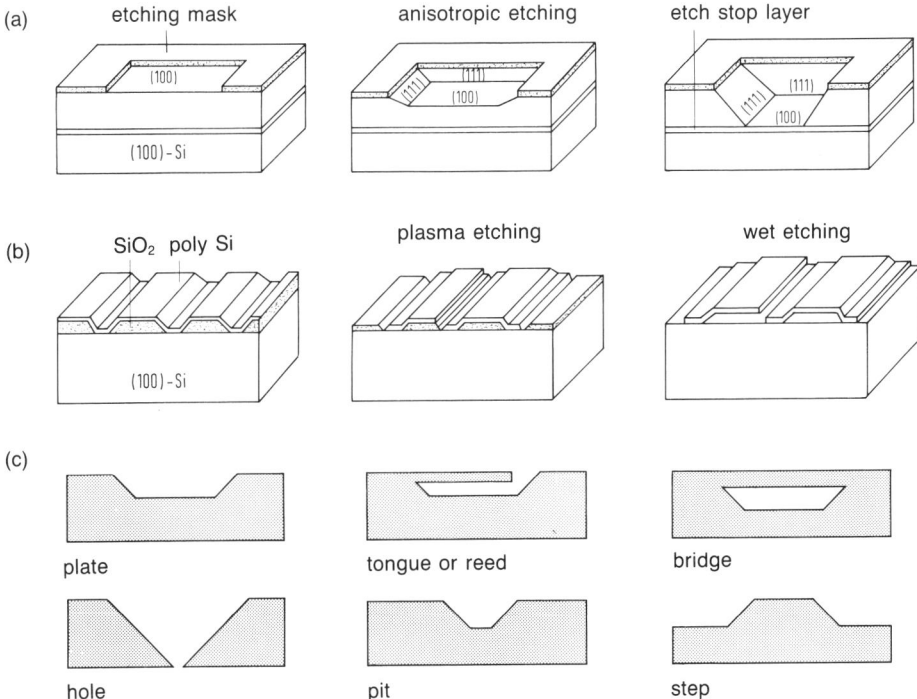

Figure 4.3 (a) Production of three-dimensional structures through the selective and anisotropic etching of monocrystalline silicon; (b) selective etching of poly-Si layers; (c) basic micromechanical structures [62].

technique, wet etching processes are combined with highly selective plasma etching processes (Figure 4.3(b)). Polycrystalline or recrystallized silicon is plasma etched and the underlying SiO_2 layer is removed by chemical etching. The lateral dimensions of such structures are larger than the thickness of the insulating layer by two orders of magnitude. This leads to the possibility of producing new structures. The main advantage of this method is that the MOS planar process can be used for the fabrication of three-dimensional structures. Plasma etching technology can be used to structure materials other than semiconductors, polymers for example. For the sake of completeness we should also mention that the techniques of micromechanics make it possible to produce micropumps, microvalves, microswitches or microloudspeakers and microphones and are therefore of interest to disciplines other than sensor technology.

4.4 Temperature sensors

Temperature is one of the most important physical dimensions. Many of the principles relating to the measurement of temperature have long been known, such as the phenomenon of mechanical expansion, the thermocouple, the metallic resistance thermometer or the pyrometer. Developments in materials sciences in the 1950s brought about additions to this list such as positive (PTC) or negative temperature coefficient (NTC) resistors. With the advances in semiconductor technology, silicon temperature sensors were also to gain considerable significance, both as discrete sensor units and as integrated, 'intelligent' sensors. In accordance with [2], we can subdivide the silicon temperature sensors into resistance temperature sensors and interface temperature sensors.

4.4.1 Resistance temperature sensors

Such sensors make use of the temperature dependence of carrier transport. Figure 4.4 depicts a sensor constructed in accordance with the principle of spreading resistance. The term 'spreading resistance' stems from a procedure for measuring

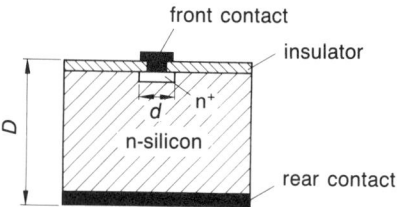

Figure 4.4 Structure of a silicon temperature sensor in accordance with the principle of spreading resistance.

the specific resistance of a semiconductor using the 'single-probe method'. The resistance R measured by a probe of diameter d is given by

$$R = \rho/2d \qquad \text{for } d \ll D \tag{4.1}$$

where ρ is the specific resistance of the substrate material. The resistance increases with temperature in accordance with the slightly curved characteristic line illustrated in Figure 4.5. Although the temperature coefficient of 8×10^{-3} K^{-1} at room temperature is higher than that of platinum, it is still lower than that of NTC resistors. However, in contrast to this latter type, silicon sensors have the advantage that they can be reproduced more reliably and with lower tolerance levels. n-Si is predominantly used in technical applications. The length of the substrate edge is 1–2 mm, the thickness approximately 200 µm. The overall diameter d has a value of 10–50 µm. A nominal resistance of 1–5 kΩ is obtained. This is a great advantage of this type of sensor. Its symmetrical construction, which uses two resistors, prevents any polarity dependence resulting from the differing contact surfaces of the front and rear contacts (Figure 4.6). The small dimensions ensure short response times. Sensors based on this principle are commercially available, for example the KTY 84 type (Valvo) for a temperature range of between $-50°$C and $300°$C.

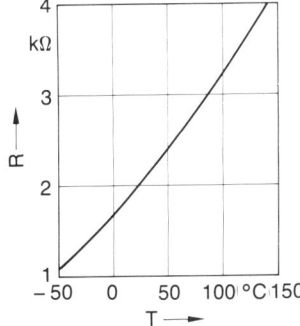

Figure 4.5 Temperature-dependent resistance of a silicon temperature sensor [63].

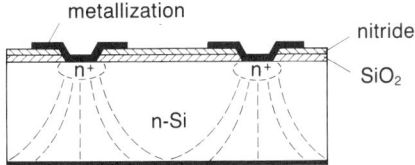

Figure 4.6 Symmetrical silicon temperature sensor in accordance with the principle of spreading resistance [4].

18 Silicon sensors

The spreading resistance sensor is a bulk component fabricated from monocrystalline silicon. However, planar resistance sensors are also usually able to make use of the temperature dependence of carrier transport. This type of sensor can be manufactured using poly-Si. Selective adjustment of the temperature coefficient across a wide range of positive and negative values (between $+2 \times 10^{-3}$ K^{-1} and -10^{-2} K^{-1}) can be achieved by varying the deposition procedure or the recrystallization parameters of the monocrystalline silicon layers. The temperature range can be extended upwards ($>200°$C) by depositing the active sensor material on an insulating substrate. The TS type sensor (Valvo) is an example of this. However, the widespread application of such sensors is only just beginning.

4.4.2 Interface temperature sensors

This type of sensor primarily exploits the temperature dependence of carrier transport by using the bias-dependent p-n junctions of diodes, transistors or transistor combinations. The effects of temperature-dependent modifications of the interface polarity of a.c.-fed MOS capacitors can also be used by this type of sensor. Both effects are used in temperature–frequency converters.

The d.c. characteristic curve of a p-n diode is strongly temperature-dependent. Thus in [31] the forward current I_F of a diode is described in terms of the forward voltage U_F as follows:

$$I_F = I_S \{\exp(eU_F/2kT) - 1\} \qquad (4.2)$$

where I_S is the saturation current. Equation (4.2) is valid when the recombination current is far greater than the diffusion current. In general, the condition $U_F \gg kT/e$ will be fulfilled, with the result that I_F and U_F are given by:

$$I_F = I_S \exp(eU_F/2kT) \qquad (4.3)$$

and

$$U_F = (2kT/e)\ln(I_F/I_S) \qquad (4.4)$$

If the ratio I_F/I_S were constant, then the result would be a sensor exhibiting ideal linear temperature-dependence of the forward voltage. However, this is not the case. I_S is dependent in a complex manner on the diode area A, the density of intrinsic conductivity n_i and temperature-dependent diffusion coefficients and diffusion lengths. Moreover, n_i is related to T via the equation

$$n_i^2 \sim \exp\{-W_g/kT\} \qquad (4.5)$$

where W_g is the energy difference in electron-volts between the valence and conduction bands. The temperature dependence of I_F and U_F is thus strongly influenced by the temperature dependence of the saturation current I_S.

A similar relationship is found in the case of transistors. If the collector and base are held at the same potential (Figure 4.7) then the relationship of the base–emitter voltage U_{BE} to the collector current I_C is given by:

$$U_{BE} = (kT/e)\ln(I_C/I_S) \tag{4.6}$$

Here again the saturation current I_S is influenced by the temperature dependence of a number of dimensions. Despite this, if the collector current I_C is held constant and the components are carefully selected, it is possible to obtain approximately linear behaviour for temperatures between $-50°C$ to $150°C$. Temperature coefficients of $2\ mV\ K^{-1}$ are achieved at room temperature for bipolar elements. In theory, the sensitivity, geometrical dimensions, price and linearity of transistors should make them the ideal temperature sensors. However, the level of temperature dependence varies greatly. Recently very large-scale integration (VLSI) techniques have made it possible to manufacture transistors with low levels of tolerance. Using this technology it is possible to produce sensors for temperatures between $-40°C$ and approximately $150°C$ with a maximum error of $\pm 2\ K$. Special calibration and compensating circuits even make it possible to reduce this temperature error to $\pm 0.1\ K$ [33]

It is relatively rare for diodes to be used as temperature sensors. In principle it is possible to obtain sensitivities of $2\ mV\ K^{-1}$ with a maximum temperature measurement error of 3% [34].

By far the most common type of temperature sensor used in practice consists of a dual transistor arrangement. In this way the problems caused by differences between individual devices can be avoided. In such a dual arrangement the difference in the base–emitter voltage is given by [31]:

$$\Delta U_{BE} = (kT/e)\ln[(I_{C2}/A_2)/(I_{C1}/A_1)] \tag{4.7}$$

The temperature dependence of U_{BE} is thus solely dependent on the ratio r of two collector current densities:

$$\Delta U_{BE} = (kT/e)\ln(r) \tag{4.8}$$

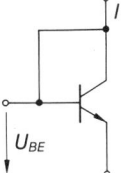

Figure 4.7 Temperature measurement using a transistor wired as a diode [4].

20 Silicon sensors

Provided that this ratio can be kept constant, ΔU_{BE} is directly proportional to the absolute temperature. There are two ways of achieving this. First, it is possible to operate two transistors with the same geometric dimensions on a single chip using two collector currents ($I_{C1} \neq I_{C2}$). The alternative is for a constant collector current to flow through two transistors with different emitter areas ($A_1 \neq A_2$). The second variant has been of greater practical relevance because of the simpler circuitry involved. An example of this type of integrated sensor is presented in the basic circuit diagram in Figure 4.8. Transistors T_1 and T_2 perform the detection function. The identical transistors T_3 and T_4 act as current mirrors. This causes a splitting of the current I into two equal collector currents I_{C1} and I_{C2}. The emitter area of T_2 should be r times that of T_1. Its collector current density is thus only $1/r$ that of T_1. The difference of ΔU_{BE} causes a current I_{C2} which is proportional to the temperature to flow across a resistor R. Because of the current mirroring, the value of I must also be proportional to the absolute temperature. Laser alignment of the resistance R makes it possible to adjust the constant of proportionality in equation (4.8) to $1 \ \mu A \ K^{-1}$. If the circuit is changed to allow for a voltage output signal then temperature coefficients of a few millivolts per kelvin can be achieved.

Commercial examples of this type of temperature sensor are the AD 590 (Analog Devices) types. They can be used to within an accuracy of approximately 1 K for temperatures between $-50°C$ and $150°C$.

Although other developments are under way, most of these are still at the laboratory stage. For example, [35] presents an integrated diode bridge as a simple, reliable temperature sensor. Sensitivity can be enhanced by increasing the number of diodes. Temperature–frequency converters represent another interesting direction of development because of their ability to provide a frequency-analog output signal. Figure 4.9 presents an example of such a sensor in the form of a temperature-sensitive ring oscillator circuit [36]. It consists of a number of inverter stages with both lateral (T_1) and vertical (T_2) transistors. The junction capacitance of the individual inverter stages causes a switching delay which, given a fixed injection current, determines the operational frequency of the ring oscillator which varies with the number of inverter stages used. The temperature dependence of U_{BE}

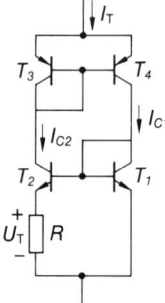

Figure 4.8 Basic block diagram of an integrated temperature sensor [4].

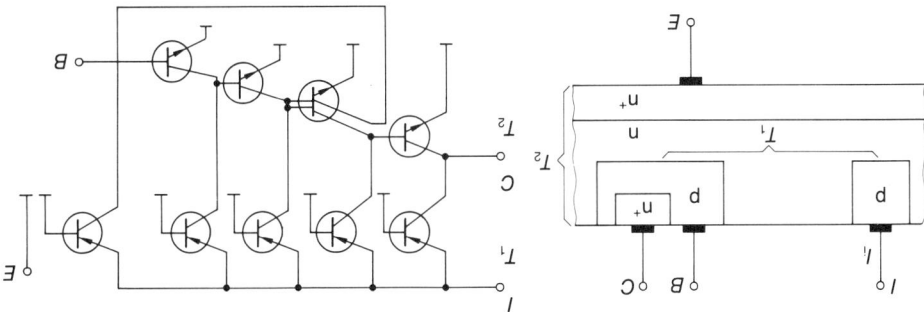

Figure 4.9 Structure and basic block diagram of a temperature–frequency converter (after [36]).

directly influences the oscillation frequency of the oscillator. For temperatures between −20°C and 80°C there is thus a linear dependence between temperature and frequency with a relative sensitivity $\Delta f/f_0$ of approximately $10^{-3}\,\mathrm{K}^{-1}$ [36]. Although the prospects for such sensors are good, they are currently unable to compete with the low-cost dual-transistor arrangements.

4.4.3 Other silicon temperature sensors and applications

At high temperatures (500°C to 3000°C), the bolometer is frequently used as a sensing element. The temperature in this device increases as a result of the absorption of thermal radiation by resistance layers. Metallic black layer resistors and metal-oxide–metal–mixed-layer resistors are frequently used. Silicon is often only used as the substrate.

Another highly sensitive radiation sensor based on thin-film technology which makes use of the thermoelectric effect is described in [37]. A thermopile structure (approximately 50 thermocouples from the active layer system $Bi_{0.9}Sb_{0.1}/Sb$) and a meandered Bi-resistor layer are applied to a silicon chip. This sensor, which has an active area of about 1 mm, exhibits a high sensitivity of 100 V/W at 25°C.

The Seebeck effect can also be used in silicon. Compared with metals, the Seebeck coefficient in silicon is high and can be modified selectively across a wide range by doping the silicon substrate. Thus there are chips with p-Si/Al thermopiles (Figure 4.10) which contain, for example, seven thermopiles and whose dimensions lie in the millimetre range. It is possible to detect temperature differences in the microkelvin range [38].

Integrated thermopiles are suitable for many applications. Alongside the direct measurement of temperature differences, they are employed in applications where the temperature is measured as an indirect dimension. Examples of such applications are flow measurements, the detection of infrared radiation and the measurement of vacuum pressures. Because silicon is a good heat conductor, etching

22 Silicon sensors

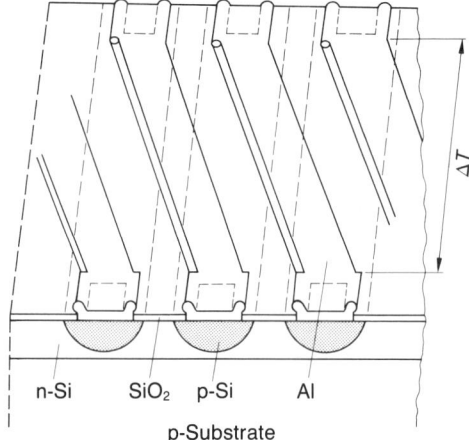

Figure 4.10 Principle of a p-Si/AL thermopile [38].

procedures can often be used to adapt the thickness and shape of the thermopiles to specific application requirements. The low offset of integrated thermopiles is a great advantage.

Amorphous silicon, with its high Seebeck coefficient, is a most promising material for use with thermopiles. At room temperature between $-120\ \mu V/K$ and $220\ \mu V/K$ can be achieved for n-material and between $170\ \mu V/K$ and $280\ \mu V/K$ for p-material. These values are better than those achieved with metals by two orders of magnitude. If a p-n junction is formed, then the sensitivity of the Seebeck effect can be increased still further. The Seebeck coefficient is approximately $300\ \mu V/K$. A p-n junction in a-Si has ohmic characteristics. The thermoelectric voltage exhibits a high degree of linearity. In this way it is possible to develop high-sensitivity temperature sensors using this class of material [39, 40].

The temperature sensors described here will form attractive alternatives to the classic resistance thermometers and thermocouples as soon as they possess smaller geometrical dimensions and can be fabricated at low cost.

4.5 Pressure sensors

Silicon sensors are of great importance in the measurement of pressure, power and acceleration. So far the sensors most commonly used for these applications have been based on the piezoresistive effect. Both high-quality/high-cost and so-called low-cost sensors have been developed. Resistances diffused or implanted in monocrystalline silicon yield the measured effect. Moreover, sensors which exploit the piezocapacitive effect are playing an increasingly important role, especially when oscillators or amplifiers can be monolithically integrated on a single chip. Piezocapacitive pressure sensors are more sensitive and stable and less susceptible to temperature changes than piezoresisitive sensors. However, the latter are simpler and cheaper to produce. In contrast to capacitive sensors, they also exhibit an approximately linear response characteristic. Signal preparation is also simpler. The latest designs include poly-Si or modified MOSFETs.

4.5.1 The piezoresistive effect

The term 'piezoresistive effect' describes the change in the electrical resistance of a material which is subjected to a mechanical stress such as tension or pressure. It occurs in crystals which have no polar axes and is well represented in semiconductors. Physically, piezoresistance comes from the anisotropic distribution of energy levels in the k-space of the angular wave vector [41, 42], a discussion of which will not form part of this book. Instead I shall present a brief phenomenological description.

In simplified scalar notation the relationship between relative changes in resistance $\Delta\rho/\rho$ (where ρ is specific resistivity) and the mechanical stress applied is given by

$$\Delta\rho/\rho = \pi\sigma \tag{4.9}$$

where π is the so-called piezoresistive coefficient which is dependent on the crystal orientation and the conditions of measurement, for example volume constancy. In reality, equation (4.9) is a tensor equation with a symmetrical tensor π containing 21 components. In the cubic system only three of the components, π_{11}, π_{12} and π_{44} are independent of each other. The same is true of monocrystalline silicon. The values of these coefficients depend on the type of conductor and the dosing level.

There are only two types of piezoresistive effect, namely longitudinal and transversal. The distinction is that in the longitudinal effect the electric current flows in the direction of the mechanical stress, whereas in the transversal variant it flows perpendicular to it. There are also two different coefficients π_L and π_T. For example, the following applies to a longitudinal change in pressure in the (111) orientation:

$$\pi_L = -\tfrac{1}{3}(\pi_{11} + 2\pi_{12}) + \tfrac{2}{3}\pi_{44} \tag{4.10}$$

Both coefficients π_L and π_T are thus highly direction-dependent [43]. The intended orientation of the piezoresistive transduction elements and their dependence on the orientation of the crystal is thus of prime importance for the technical applications for which the sensor can be used. The piezoresistive coefficients for the usual orientations of piezoresistive transduction elements are extensively detailed in [44].

In practice, piezoresistance is mostly used for elements which are applied to a deforming body in the form of a resistance circuit. The deforming body usually takes the form of a flexible bar, especially in power and acceleration sensors, or of

24 Silicon sensors

a rectangular or circular diaphragm. The resistors are bridge connected in the regions of maximum mechanical stress. Classically, a so-called wire strain gauge (WSG), made of metal, constantan for example, is affixed to the deforming body. The bridge connection requires four resistors with characteristics which are as similar as possible. This is ideally realized by means of microelectronic techniques in which four resistances are diffused in a semiconductor of a given doping level, for example monocrystalline silicon. This method possesses a second advantage which can be explained by the following observation. Given an idealized rod of length l which is subjected to a mechanical stress σ then Hooke's law applies:

$$\sigma = E(\Delta l/l) \tag{4.11}$$

where E represents the elasticity and Δl the change in length under tension. Equation (4.9) yields the following relative resistance change in the longitudinal resistance effect:

$$\Delta R/R = k(\Delta l/l) \tag{4.12}$$

with the new constant of proportionality $k = \pi_L E$ (at constant volume). The constant is known as the *k factor*. It is dependent on the material used and in silicon is not a function of the doping level alone, but also of the orientation of the crystal. In silicon k factors of up to 180 can be achieved. A typical value is 120. In contrast, metals have k factors ranging from only 2 (constantan) to 6.6 (Pt–Ir). In these the relative resistance change $\Delta R/R$ is given by the geometrical change (Δl, Δq) (for cross-section q) alone. $\Delta \rho/\rho$ is approximately zero.

However, in semiconductors ρ changes greatly when the material is deformed. Silicon resistors are therefore much more sensitive to changes in shape than metal ones. The sign of the k factor depends on the type of conductor and the doping level ($k > 0$ for p-conductors and $k < 0$ for n-conductors) and its value depends on the crystal orientation. For all orientations, the absolute value of k increases with resistivity. Unfortunately the same is true of the sensitivity to temperature changes.

4.5.2 Piezoresistive pressure sensors

The measuring element of a monolithic integrated piezoresistive pressure sensor is illustrated in Figure 4.11 as an example of the use of the piezoresistive effect. A membrane is etched on a silicon substrate. The thickness of the membrane can vary from a few micrometres to millimetres depending on the pressure to be measured. The membrane functions as a built-in diaphragm. When it is deformed, its surface is stretched and compressed. At this point resistances are integrated by means of diffusion or ion implantation and these are correspondingly stretched or compressed. The actual arrangement of the elements on the membrane depends on the orientation of the crystal (Figure 4.12). Together with the piezoresistive elements, the simplest configurations of piezoresistive sensors, which have been

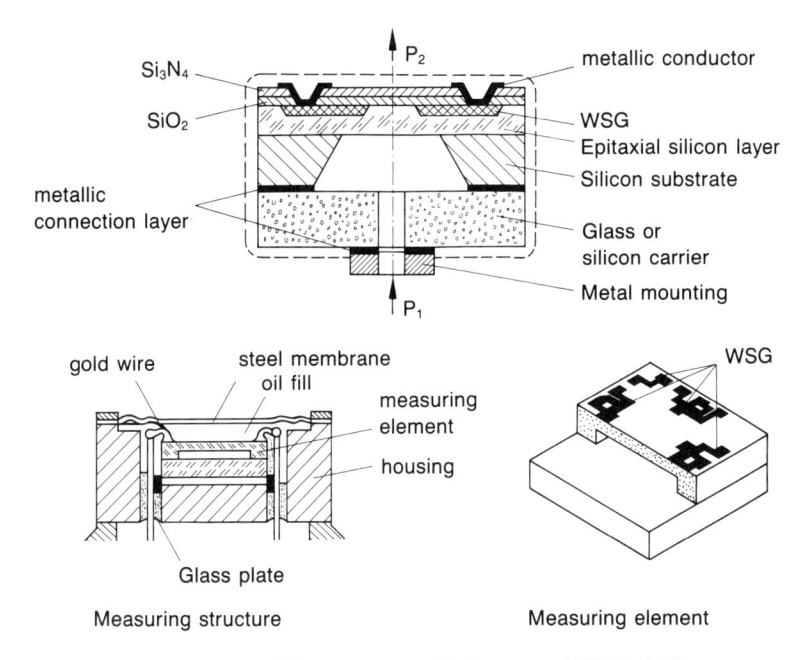

Figure 4.11 Silicon sensor with integrated WSG [47].

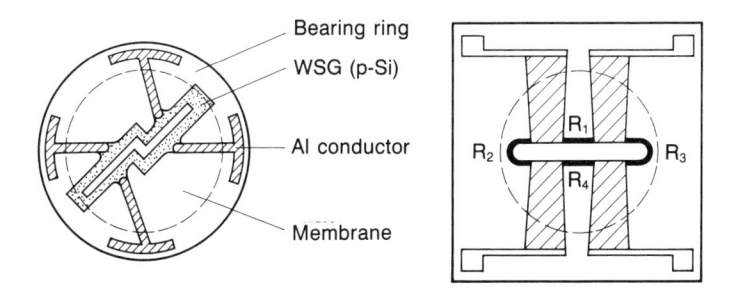

Figure 4.12 Example of an arrangement of a semiconductor WSG on a pressure-sensitive membrane.

manufactured for a number of years, possess resistors for balancing the bridge voltage (offset), for balancing the sensitivity of the sensor and for temperature measurement. This compensates for manufacturing tolerances such as imbalances in the bridge resistances, membrane and material tolerances or temperature dependences. The effect of temperature manifests itself in, for example, the temperature coefficient of resistance (TCR). A measure of the susceptibility of the sensor to temperature gradients is given by the ratio of the TCR to the full bridge deflection. Using a metal WSG, TCR values of $10^{-5}\,\text{K}^{-1}$ at a full deflection of

26 Silicon sensors

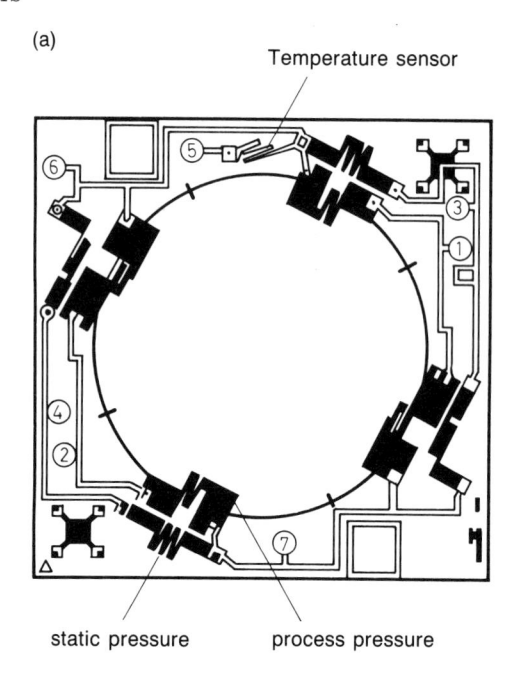

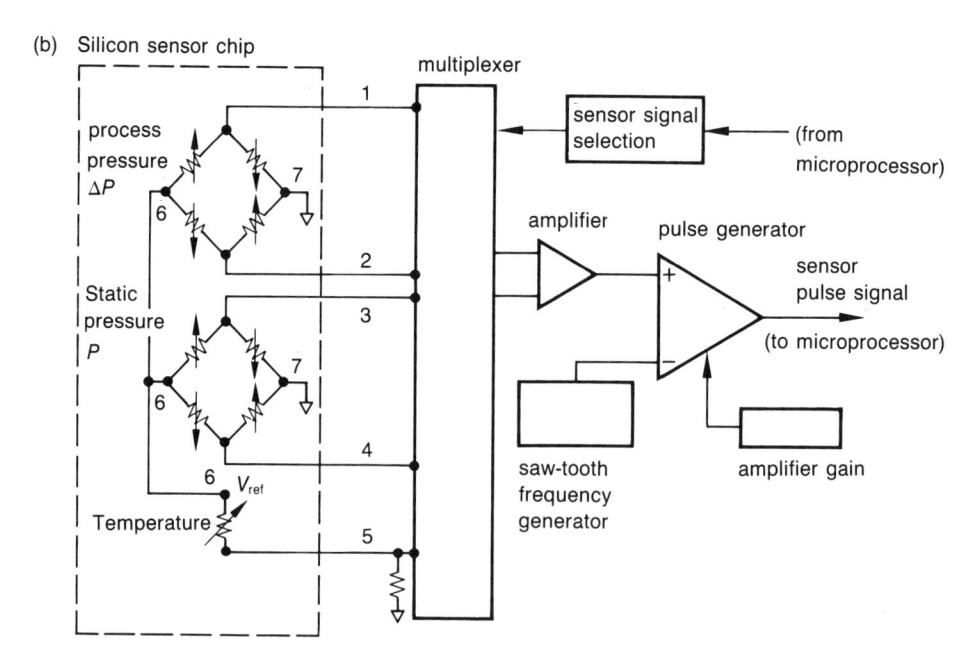

Figure 4.13 (a) Sensor element of the ST 3000; (b) diagram of the sensor element and the principle of generation of a pulse-width modulated digital signal ready for further processing.

2 mV/V are achieved. Semiconductor-based sensors have a TCR of 2×10^{-3} K^{-1} at an output signal of 20 mV/V and are therefore worse by a factor of 20. The uncorrected value of the temperature coefficient (TC) of the sensor sensitivity, which is composed of the TC of the sensitivity to extension (which is determined by the k factor) and the temperature variation of the modulus of elasticity of the silicon membrane, is 2×10^{-3} K^{-1}. Both examples illustrate the importance of balancing the temperature of semiconductor pressure sensors. In modern systems this is achieved by means of laser trimming. Computer-aided laser cutting systems are used to perform both coarse and fine alignments, for example at metal thin-film balancing resistors. This considerably reduces the influence of temperature and linearity errors.

Pressure sensors based on monocrystalline silicon are well suited to many applications which do not require an accuracy in excess of 0.5%. Many manufacturers now produce sensors for nominal pressures of 1 mbar to 1000 bar. These are suitable for both absolute and differential pressure measurements. They are not susceptible to overloads. However, they break easily when the maximum permitted pressure is exceeded. They should be protected from aggressive materials. For use in industrial applications the chip is sealed in a hermetic housing with a cellular metal membrane and is embedded in an oil film. The maximum operating temperature is approximately 120°C since at higher temperatures the insulation between, for example, the p-doped WSG and the n-doped spring element deteriorates too greatly. Many producers are working on improving the accuracy of this type of sensor and on reducing the lower nominal pressure. Both developments place considerable technological demands on the manufacturer.

Alongside the sensor itself, more demanding pressure sensor configurations include electronics for signal preparation. These can either be separate or take a monolithically integrated form. A well-known example of the discrete variant is the first so-called 'smart' pressure sensor, the ST3000 (Honeywell) [45]. The skilful arrangement of resistor combinations on the chip and microprocessor-controlled preparation of the pulse-width modulated sensor signal gives this sensor noteworthy characteristics (Figure 4.13). The most important of these is the wide measuring range, with a ratio of 1:400. An integrated variant is illustrated in Figure 4.14. In this case integration is performed using bipolar technology [46]. Integration brings a number of advantages. Laser alignment is performed during production, a vigorous output signal is obtained (up to 90% of the operating voltage) and inter-manufacturer replacements can be used.

Frequency-analog pressure sensors can also be produced by integrating oscillator circuits into the chip. If a ring oscillator is used for the converter circuit in a piezoresistive sensor, the injection current and thus the frequency of the oscillator can be controlled by modifying the resistance [36, 48, 49]. If the ring oscillators are connected in series, as illustrated in Figure 4.15 the following linear relationship between resistance change and frequency shift is obtained:

$$R_\text{L} - R_\text{Q} = 2\Delta R = c(f_\text{L} - f_\text{Q}) \tag{4.13}$$

28 Silicon sensors

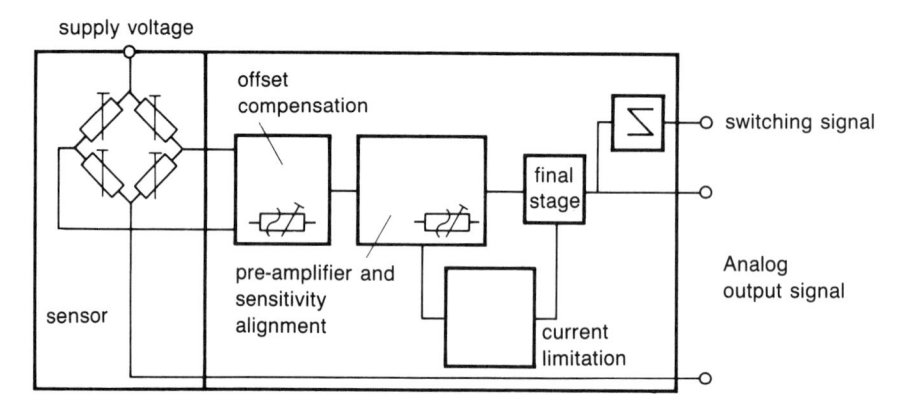

Figure 4.14 Block diagram of the integrated signal preparation in a pressure sensor [46].

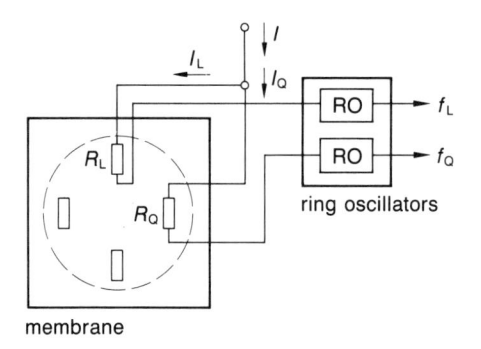

Figure 4.15 Frequency-analog pressure sensor [36].

If a nominal frequency of 200 kHz is prespecified a change in frequency of 30 kHz can be obtained at a pressure change of $0-10^5$ Pa. Sensors of this type can be used within a temperature range of $-30°C$ to $+80°C$. However, the high degree of temperature dependence of the injection current means that they have a relatively high temperature coefficient, greater than 10^{-3} K^{-1}. This can be reduced by using a frequency response ratio as the output signal. A presentation of a sensor which makes use of a two-stage ring oscillator, each stage of which consists of nine-stage MOSFET ring oscillators, can be found in [50]. The pressure sensitivity of the carrier mobility in the depletion channel is used as the pressure-sensitive element. Since then there have been many developments in this direction [51–54]. The most recent work has been conducted with the aim of realizing a semicustom design for this type of frequency-analog pressure sensor [55].

A pressure sensor with a digital output signal was recently presented as a 'statistical' NMOS flip-flop sensor [56]. Figure 4.16 depicts the structure of this type of sensor. Initially the flip-flop structure is in an unstable state. This state can be

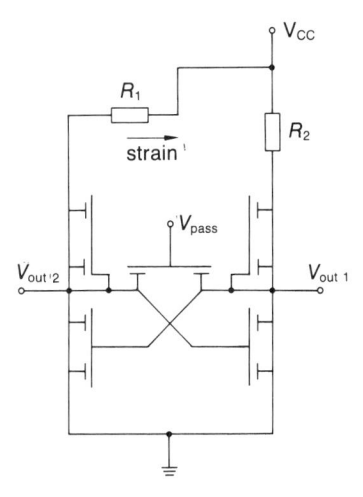

Figure 4.16 Flip-flop structure with two additional piezoresistors R_1 and R_2 [57].

influenced by two piezoresistive resistances which are affected by the pressure measurand. This results in a change in the initially random pulse train at the sensor output. It is possible to design extremely sensitive sensors based on this principle. Interestingly, the same operating principle can be found in optical, magnetic and temperature sensors [57]. However, this work is still in its infancy.

From the technological point of view, the poly-Si pressure sensor is a relatively new and interesting pressure sensor. It is manufactured using silicon on isolator (SOI) technology. In principle this is a thin-film technology, which will be discussed later. In this technology the sensitive element is electrically isolated from the silicon substrate. This yields many advantages. Temperature- and time-unstable p-n junctions are avoided. The pressure sensor is thus more stable over longer periods, its operating temperature range extends up to 200°C and the temperature drift of

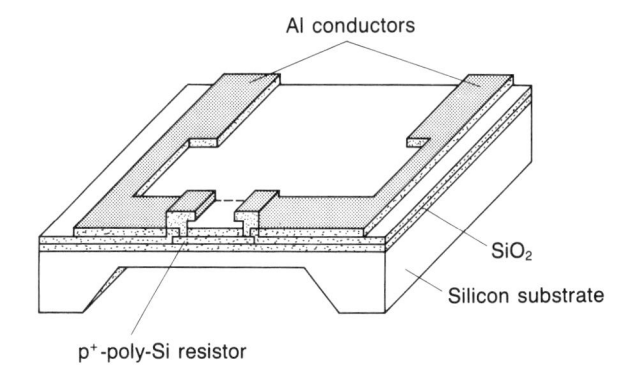

Figure 4.17 Piezoresistive sensor with poly-Si resistors [62].

the output voltage can be considerably reduced in comparison to monocrystalline silicon. The first commercial sensors appeared in 1988. Figure 4.17 illustrates the structure of this type of sensor. The piezoresistive sensor material is poly-Si. Its properties can be selectively modified by altering the deposition conditions and subsequent treatment. This means that relatively low k factors (between 10 and 25) are obtained for plasma CVD silicon layers [59]. The low processing temperature means that deposition is possible on the most varied spring materials from metals to plastics (Figure 4.18). This is a great advantage. It is possible to adapt such sensors to a great variety of measurement tasks. It is possible to achieve higher k factors by tempering the poly-Si. However, if this is done then the advantage mentioned above is lost. Zone melting processes can then be used to produce recrystallized silicon layers with piezoresistive properties similar to those of monocrystalline silicon.

4.5.3 Capacitive pressure sensors

The capacitance C of a plate capacitor depends on the relative dielectric constant ε_r, the distance d between the plates and the plate area A:

$$C = \varepsilon_0 \varepsilon_r A / d \tag{4.14}$$

Changes in ε_r, A and d are mainly of use in sensor applications. For pressure sensors it is primarily the change in capacitance as a result of the deformation of a membrane, that is to say, the change in distance Δd, that is used to provide the sensing effect. Particularly small capacitive sensors can be produced using silicon technology. For example, a capacitor plate can consist of a membrane which remains after the two sides of the silicon substrate have been etched (Figure 4.19). Opposite this there is a partially metalled glass plate. The metal film forms the

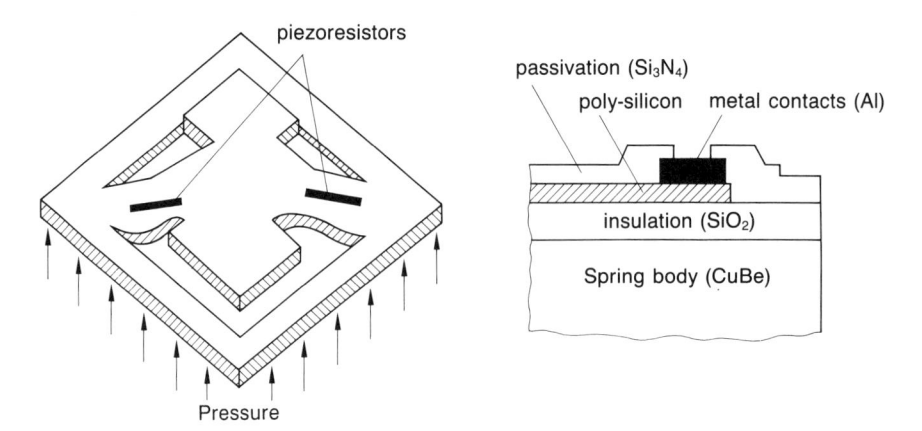

Figure 4.18 Polycrystalline silicon based pressure sensor with metal spring body [58].

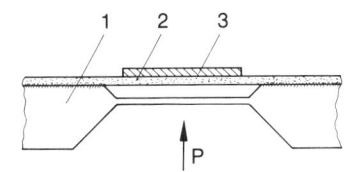

Figure 4.19 Principle of a capacitive pressure sensor (1: Si; 2: glass plate; 3: metallization).

second electrode of the capacitor. The glass and the silicon are connected by anodic bonding. Very small changes in capacitance occur, and these are difficult to detect electronically. For this reason it is preferable for the processing electronics to be integrated on the silicon substrate, for example in the form of an RC oscillator. The variable capacitance of the oscillator then becomes the pressure sensing mechanism. A periodic signal occurs at the output. The latest microconstruction techniques make use of silicon on silicon bonds [59]. This means that the sensor can only be made of silicon wafers (Figure 4.20). Layers 1 and 2 form the capacitor plate. Levels of sensitivity of 0.1 fF/mbar can be achieved. The oscillator circuit is integrated using CMOS technology. The batch processing of silicon-metal structures is also possible using microstructuring methods [61]. It can be assumed that capacitive silicon sensors will grow in importance in the future.

4.5.4 New pressure sensor principles

The integration of a MOSFET and an electret has led to the development of a wide variety of new sensors, a number of which are pressure sensors. For example, the new designation PRESSFET has been given to a pressure sensor whose structure is illustrated diagrammatically in Figure 4.21 [64–66]. It consists of a thin, conductive diaphragm, an air gap, the electret (teflon is used in this particular sensor), an SiO_2

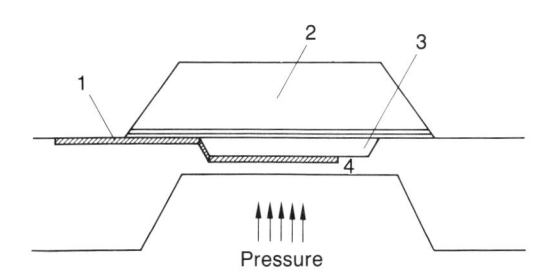

Figure 4.20 Integrated silicon-based pressure sensor [59].
1 Layer of diffused boron
2 Silicon cover with diffused phosphorus
3 Reference cavity
4 Diaphragm

32 Silicon sensors

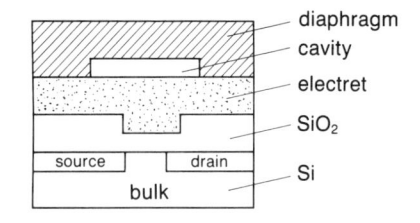

Figure 4.21 Schematic representation of a PRESSFET [66].

layer and the silicon substrate. The PRESSFET can be considered as a new type of FET arrangement with a dielectric sandwich layer between the gate (the conductive diaphragm) and the silicon. The current in the drain–source channel depends, among other things, on the capacitance between the gate and the silicon. It is thus a function of the applied pressure. Reactive pressure in the range 20–400 kPa can be detected.

A silicon reed oscillator, manufactured using micromechanical methods, can be used as a vacuum sensor. A silicon paddle whose dimensions lie in the millimetre range (Figure 4.22), and which has been caused to oscillate, exhibits a pressure-dependent change in the amplitude of its vibrations. This principle can be used for measuring pressures of between 10^{-2} and 10^5 Pa.

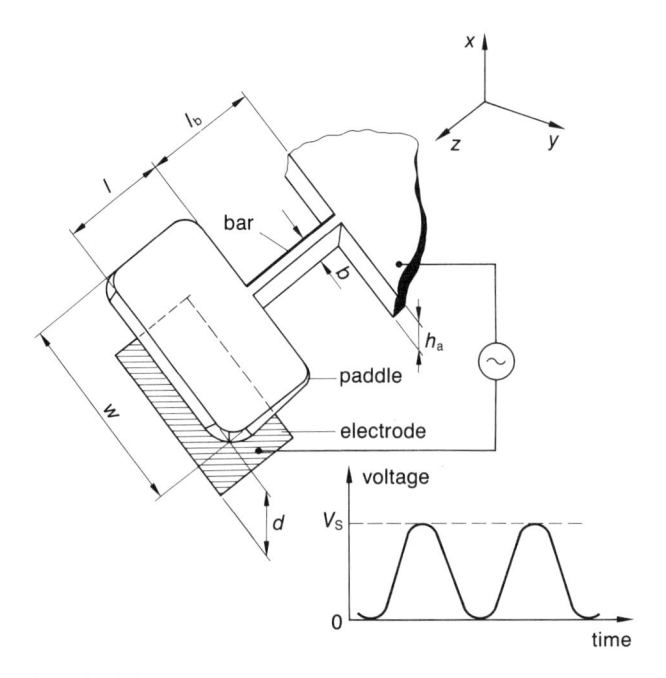

Figure 4.22 The principle underlying a silicon reed oscillator used as a pressure sensor.

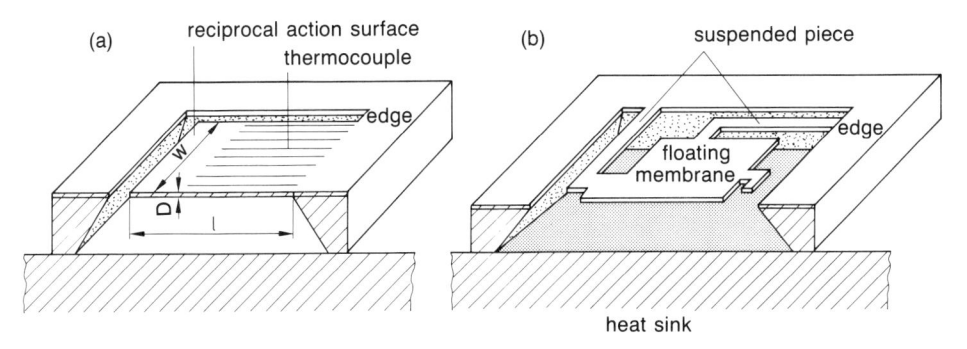

Figure 4.23 Thermal vacuum sensors [68]: (a) reed sensor ($w = 1.7$ mm, $l = 6$ mm, $D = 10$ μm); (b) floating membrane sensor (thickness of membrane: 10 μm, area 3.4×3.4 mm).

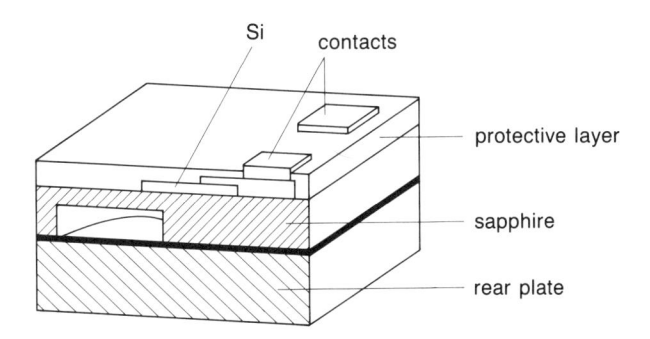

Figure 4.24 SOS pressure sensor.

Sensors which are also produced for vacuum measurement using micromechanical methods but which are based on thermal principles are presented in [68]. Their operating principles are similar to those employed in the Pirani vacuum gauge. This means that the vacuum pressure is deduced from the measured change in the thermal conductivity of the gas. Two particularly suitable principles have been developed, the thermal reed sensor and the integrated floating membrane sensor (Figure 4.23). In the first type, an integrated thermopile is located on the reed or cantilever beam (see Figure 4.10). Such sensors can be used for pressures between 10 mPa and 10 Pa. In the sensor illustrated in Figure 4.23(b), thermopiles are located on the bars of the suspended piece, a floating membrane. These thermopiles measure the increase in the membrane temperature relative to the surrounding temperature. The temperature increase is generated by resistors located at the junction of the bars and the membrane. The electrical power necessary to maintain a constant temperature in the presence of a cooling gas is provided by a resistor which covers the entire membrane. The power is linearly dependent on the pressure p. It is possible to measure a pressure range from 10 mPa to 10 kPa.

In many industrial or laboratory applications it is common to encounter temperatures higher than 100°C. It is then necessary to use piezoresistive sensors which are produced using silicon on sapphire (SOS) technology. Such SOS pressure sensors consist of two parts, the sapphire (Al_2O_3) containing the diaphragm and the resistors, and the base, which may be made of glass for example (Figure 4.24). These sensors can be used at temperatures of up to 425°C and pressures of up to 10 Pa [69].

4.6 Optical sensors

Semiconductor-based optical sensors are of great importance in the fields of measuring and automation technology. However, optical sensors are only infrequently used to measure light itself. Instead, they are generally employed as tools for the measurement of other quantities such as position or path of travel, with light playing the role of information transport medium in many sensors. The most important criteria for the industrial use of sensors are the sensors' universality, their simplicity of use and their compatibility with microelectronic equipment. This is the reason why nowadays silicon sensors are almost exclusively used. Optical semiconductor sensors detect the presence of carriers generated in the bulk of the sensor by the interaction between light and the semiconductor. The free carriers are generated by ionization. This process requires a certain amount of energy to be absorbed by the crystal. In an ideal semiconductor the only absorption process which is possible at low temperatures consists of the excitation of bound valence electrons. The required photon energy must equal or exceed the value of the bandgap E_g (Figure 4.25(a)), i.e.

$$E_{ph} = h\nu = hc/\lambda = E_g \tag{4.15}$$

For Si, the value of E_g at room temperature is 1.12 eV. However, in a real semiconductor crystal other excitation mechanisms are possible. These include absorption through transitions between the allowed bands and absorption through high levels of distortion in the forbidden band. However, the greatest excitation effect is still provoked by so-called fundamental absorption, that is to say, direct band-to-band absorption.

Alongside the 'internal' photoelectric effect (photoconductivity) in semiconductors, a particular form of the 'external' photoelectric effect (photoemission) which occurs at metal–semiconductor junctions has a role to play in optical sensors. If a metal is brought into contact with a semiconductor, then the movement of charge which occurs at the interface may lead to the creation of a depletion region. This effect is caused by the different ionization energies of the two materials. In the case illustrated in Figure 4.25(b), which depicts n-Si with an ionization energy of Φ_{SC} and a metal with an ionization energy of Φ_M, thermal contact at thermal

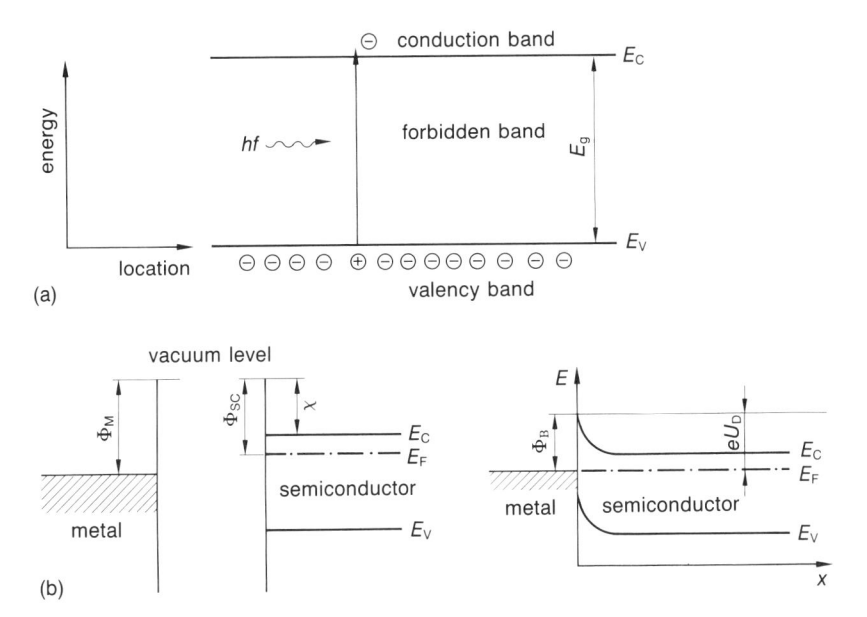

Figure 4.25 (a) The generation of an electron–hole pair through optical absorption schematically illustrated by means of a semiconductor band model; (b) band model for the ideal contact between metal and semiconductor in a Schottky junction [70].

equilibrium causes a band distortion in the semiconductor. The value of this distortion equals the difference between the ionization energies

$$eU_D = \Phi_M - \Phi_{SC} \tag{4.16}$$

A potential barrier with the value Φ_B exists between the metal and the semiconductor. If such a structure is subjected to photon radiation with an energy $hf > \Phi_B$, then metal electrons are raised to a level which enables them to cross the barrier and appear as excess carriers in the semiconductor. In this way it is possible to shift the sensitivity range, which is limited by E_g, to higher wavelengths.

It is not possible to describe a-Si and poly-Si using the energy-band model [71]. Amorphous silicon is characterized by a high optical absorption coefficient and a high photoconductive capacity. Criteria of efficiency such as spectral sensitivity, absolute sensitivity, response time and the signal-to-noise ratio have been introduced to evaluate the characteristics of optical sensors. A detailed discussion of these can be found in [70]. As explained in [2], we can differentiate between the following basic groups of components and operating principles:

1. bulk-controlled photosensors such as photoresistors;
2. interface-controlled photosensors with a single interface such as a p-n or MS junction or with multiple interfaces such as an n-p-n or p-n-p-n junction.

36 Silicon sensors

In this section photoresistors, photodiodes and transistors, diode arrays and CCD elements will be presented.

4.6.1 Photoresistors

Photoresistors are photoconductors constructed of materials whose optical conductivity changes when they are exposed to light. A well-known example is provided by the thin-film photoresistors consisting of polycrystalline CdS or CdS/CdSe heterojunction semiconductors. However, advances in Si technology mean that these have been superseded by Si components, for example by Si photodiodes connected to form photoresistors.

Pure monocrystalline Si has been largely ignored as a photoresistor material on account of its relatively low level of resistance change. More useful characteristics are exhibited by a-Si:H. Photoresistors made from this material are fabricated on insulating substrates in both lateral and vertical configurations. Figure 4.26 illustrates both alternatives. In both cases we are dealing with a thin-film resistor with a thickness of 0.5–1 µm. In Figure 4.26(b) this has been bonded with a transparent, conductive indium-tin-oxide (ITO) layer. The response time lies in the millisecond range. Such devices provide an approximately linear correlation between incident light and photoconductive capacity.

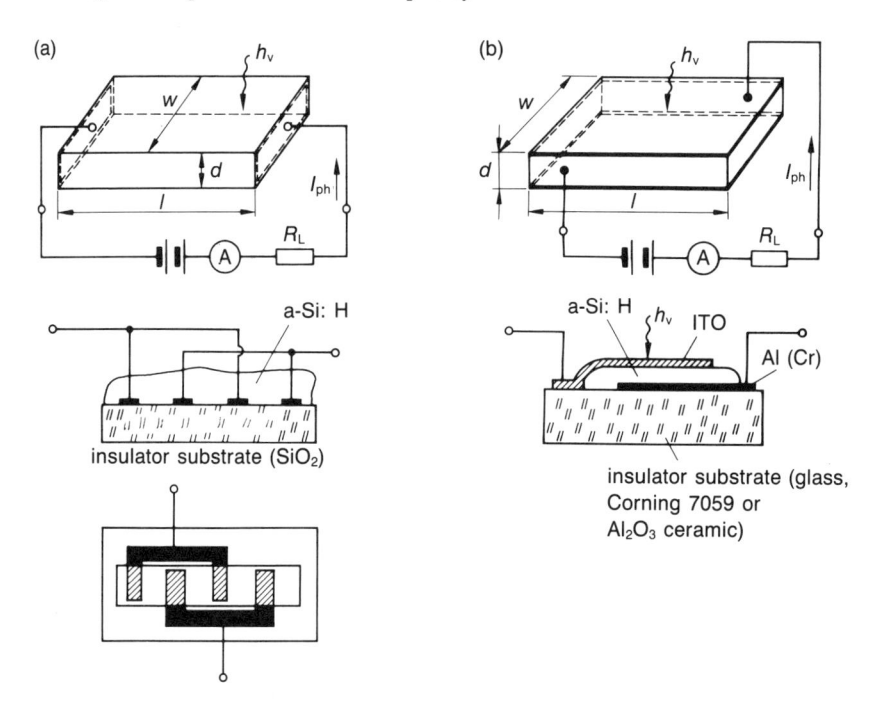

Figure 4.26 Design forms of photoresistors made of a-Si:H [2]: (a) lateral photoresistor; (b) vertical photoresistor.

4.6.2 Photodiodes and phototransistors

Optical sensors predominantly make use of Si detectors with p-n junctions. These can be used as photoelectric elements without any external voltage or as diodes in the presence of a reverse bias (Figure 4.27). In p-n photodiodes the light is mostly absorbed in the upper p-layer. The minority carriers which this creates diffuse into the depletion region where they are trapped by the built-in field. The photocurrent is, at a remove of several orders of magnitude, a linear function of the light energy falling on the light-sensitive surface. The maximum spectral sensitivity is reached at approximately 850 nm. The response times vary depending on the active surface and lie in the nanosecond range. By applying special antireflection coatings to the surface, it is possible to shift the maximum sensitivity to shorter wavelengths of 0.4–0.6 µm.

The more closely the value of the radiation energy approaches the bandgap, the greater the penetration depth of the photons and, in consequence, the greater the collection volume – the field-filled zone of the diode – must be. In an undoped (intrinsic or i-) semiconductor there are no depletion layers. If there is an intrinsic zone between the diode's p- and n-zones (Figure 4.28), then the reverse bias is reduced in this high-resistance zone, producing a constant field which separates the optically generated electron–hole pair. This makes the collector region bias-independent and means that it can be optically configured during production for a desired cutoff frequency at a particular wavelength. This ability makes p-i-n photodiodes extremely attractive for a number of applications. In a Schottky diode, photons with energy $hf > E_g$ produce electron–hole pairs in the depletion region of the Schottky barrier. They are separated by the field in a way analogous to the photoelectric effect at the p-n junction. The 'external' photoelectric effect, in which electrons are emitted from the metal to the semiconductor, is smaller than the band-to-band excitation effect by approximately 2 orders of magnitude (Figure 4.29). Because of their surface barriers, Schottky diodes

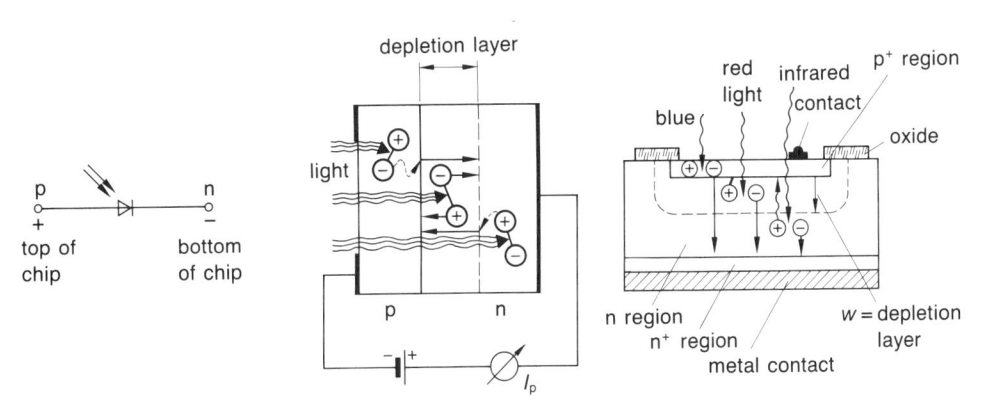

Figure 4.27 Construction and operating principles of a planar photodiode.

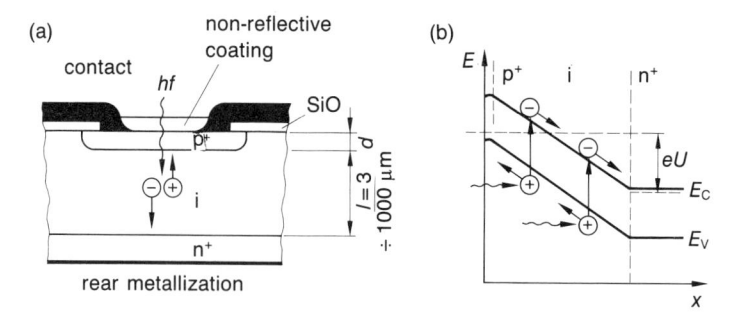

Figure 4.28 Diagrammatic construction (a) of a p-i-n photodiode; (b) and the associated band model [70].

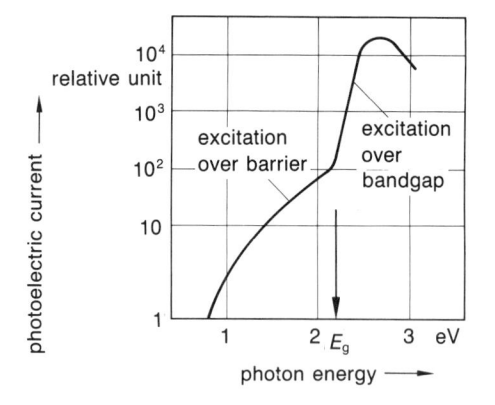

Figure 4.29 Photoelectric current of a Schottky diode as a function of the photon energy [70].

(Figure 4.30) are well suited to detecting UV radiation which has a penetration depth of the order of 0.1 μm. Schottky diodes sensitive to $hf < E_g$ are extremely fast because the time constant is determined only by the time the electrons require to cross the metal–semiconductor interface. This allows frequencies of up to 20 GHz to be reached.

If the energy of the radiation to be detected is very low, then the photoelectric current which is to be generated may be smaller than the dark current. In such cases increased current sensitivity can be obtained by internal carrier multiplication such as is found in avalanche photodiodes. However, this requires high field strengths, close to the breakdown voltage, at the p-n junction. This complicates the construction of these diodes and means that exacting technical requirements must be satisfied during production. The simple photodiode can be extended to form a bipolar phototransistor (Figure 4.31). This makes it possible to amplify the photoelectric current by a factor of between 100 and 1000, and this increases the sensitivity of the sensor. However, phototransistors have rise and fall times of

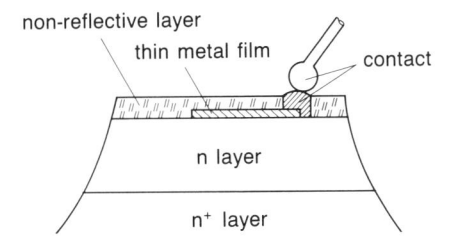

Figure 4.30 Diagrammatic construction of a Schottky photodiode [70].

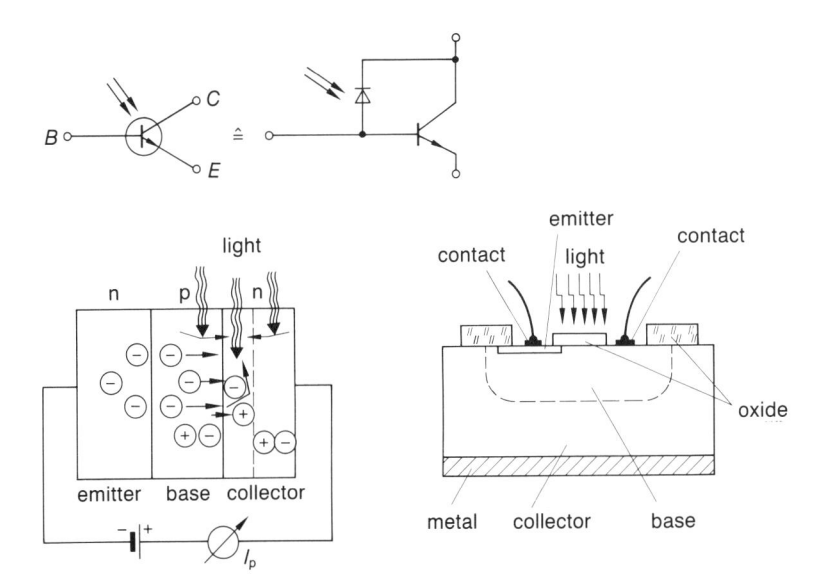

Figure 4.31 Construction and operating principle of a phototransistor.

5–10 µm and are thus much slower than photodiodes. If, for example, the collector–base junction of an n-p-n transistor operating in reverse direction (Figure 4.31) is illuminated, then the optically generated carriers generate a collector–base current I_{CB}. This in turn causes a change in the emitter current. The collector–emitter current I_{CE}

$$I_{CE} = I_{CB}(1 + \beta) \qquad (4.17)$$

is measured, where β represents the current amplification. The transistor provides an output current which is proportional to the incident radiation as long as β is current-independent. However, this factor varies with different sensors and is temperature-dependent, with the result that there are usually deviations from linearity.

Alongside phototransistors, photothyristors and photo-FETs are other well-known multi-junction components for special applications [70]. However, they have not attained the same importance as the sensors mentioned earlier. Photo-FETs are of interest in connection with UV radiation.

Nowadays the photosensitive sensors listed above are manufactured and marketed in a wide variety of types by most producers of semiconductors. An example of this is the wide range of SP type sensors manufactured by WF Berlin. The objective of the most recent developments is to integrate the sensor using a cheap integration or hybrid technique.

The traditional photodiode is not sensitive to the colour of the incident light. The resulting photocurrent depends on the number of photons absorbed by the semiconductor. Different photon energies cannot be detected. The diffusion length is much longer than the width of the depletion layer. This type of diode could therefore be designated as a diffusion type photodiode. If different colours have to be detected, then it is necessary to use optical filters. The result is that colour detection is an expensive task. Despite this, the significance of colour sensors is increasing dramatically as an element in the field of flexible automation. Here, a completely new optical sensor has entered the market, the so-called drift type photodiode [73, 74]. In this sensor the lifetime of the minority carriers is so short that only the carriers which are generated in the depletion zone of the junction can be collected by the electrons. The excess carriers which are generated outside the depletion zone, that is to say, in the field-free area, recombine before they diffuse to the junction. This means that they are not collected and therefore do not contribute to the photocurrent. In this type of photodiode the photocurrent increases with the reverse bias, that is to say, as the width of the depletion layers increases. This effect varies with the wavelength. In the blue–green range the optical absorption coefficient is high and in consequence the penetration depth is small. The photocurrent rapidly reaches saturation as the reverse bias increases. In the yellow–red range the absorption coefficient is low. As the reverse bias increases, the photocurrent moves slowly towards saturation. This means that spectral sensitivity is a function of both the reverse bias and the wavelength of the incident light. Generally speaking, the following holds for the photocurrent for this type of diode:

$$I(U) = \int_0^\infty F(\lambda) A(\lambda, U) \, d\lambda \tag{4.18}$$

where U is the reverse bias, $F(\lambda)$ the energy spectrum of the incident light, and $A(\lambda, U)$ the spectral sensitivity of the photodiode. Equation (4.18) shows that $F(\lambda)$ can be obtained as an inverse integral transformation, provided that $A(\lambda, U)$ has already been defined. It is precisely this principle that is now being applied in various arrangements of photodiodes [73–75]. Amorphous silicon is an important base material in such photodiodes because of its low carrier mobility. As an example, Figure 4.32 depicts a structure which makes use of a Schottky barrier. The diffusion length of the minority carriers in undoped a-Si:H (i layer) is 0.1 µm and is thus shorter than the width of the depletion layer. Given this arrangement, it is then

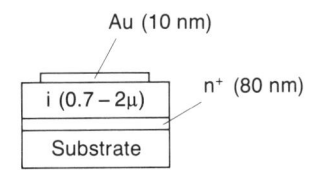

Figure 4.32 Amorphous silicon based colour-sensitive photosensor [75] (substrate: steel or chrome-plated glass).

possible to measure both the light intensity and specific colour components, for example four in [75], in the visible range of the incident light simply by varying the reverse bias. The design of such sensors demands the greatest precision.

Another advantageous variant of sensors with spectral sensitivity employs vertical arrangements of photosensitive semiconductor layers. If n wavelengths are to be detected then $n+1$ layers must be present. The photosensitive layers are separated by isolating or electrically conductive layers. The first structures of this type, for example in the form of vertically arranged p-n junctions, appeared many years ago. a-Si:H can again be used as the photosensitive material. Figure 4.33 depicts two designs for vertically arranged components based on a-Si:H [76]. The photoelectric signals S_1 (caused by the part of the light radiation which is absorbed in the first semiconductor layer) and the signal S_2 (caused by the light energy absorbed by the second semiconductor layer) are defined. It can be shown that the

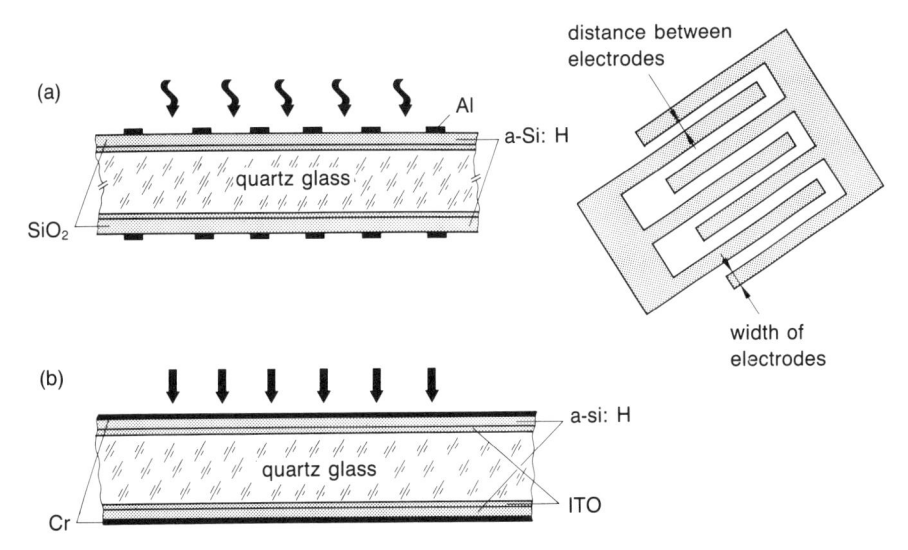

Figure 4.33 Example of the design of a spectrally sensitive sensor with vertically arranged photosensitive semiconductor layers [76]: (a) sensor with lateral resistor geometry; (b) sensor with vertical resistor geometry.

wavelength of monochromatic light can be precisely determined from the quotient S_1/S_2. Independently of this, the intensity can be determined from either S_1 or S_2. In the case of polychromatic radiation it is not the wavelength but the focus of the spectral distribution that is determined. This new spectrum-sensitive sensor can be employed in a number of applications, for example colour identification, for determining differences in colours, for measuring the thicknesses of layers, etc. Further technological development, such as the application of more than two a-Si:H layers, varying the thickness of the layers or the production of arrays should make more interesting areas of application available in the future.

In many applications it is necessary to determine the position at which the radiation strikes the photosensitive sensor. This requires the use of position-sensitive photodiodes. So far we have assumed that the whole of the photosensitive area is uniformly illuminated. Position-sensitive elements make use of the lateral photoelectric effect [2]. This is based on the fact that when a junction (p-n junction, Schottky or MOS barrier) is non-uniformly illuminated, a lateral photoelectric potential is generated. If the boundary potentials are held constant the corresponding lateral photoelectric current is a linear function of the x or y coordinates. This means that it is possible to measure the position of the incident light. These sensors are referred to as *position-sensitive full-area* photodiodes. They have a relatively large active area (25–1000 mm^2). The lateral homogeneity of the junction must satisfy exacting requirements. Special design forms lead to one-dimensional diodes which are a few millimetres in width and which make it possible to determine the position of a light dot to an accuracy of 10^{-4}.

Position measurements can also be performed by photosensors in which multiple diodes are integrated. The position is displayed by means of the geometrical distribution of the incident light pencil over a number of photodiodes. In their simplest combination these diodes are known as *differential* or *quadrant* photodiodes. They are manufactured in large quantities for commercial use. Besides these simple combinations there are also monolithic integrated detector circuits in which numerous receivers are arranged either in sequence or distributed across a surface. These sensors, which can be used to evaluate lengths or images, are the photodiode arrays and CCD elements which will be presented in the following.

4.6.3 Photodiode arrays

These are self-reading solid-state image sensors. They are based on the principle of integrating the photoelectrically generated carriers in the depletion layer of a p-n junction. This type of image sensor is frequently known as a *photodiode MOS array* since the photosensitive diodes are read via MOSFETs. The photosensitive diodes are monolithically integrated with the reading electronics on a silicon substrate. Each cell (Figure 4.34) consists of a large-area p-n junction – with storage capacity C_S – of which a small area is light-sensitive thanks to the presence of a vapour-deposited mask. Each capacitor is connected to the video output via a MOSFET.

Optical sensors 43

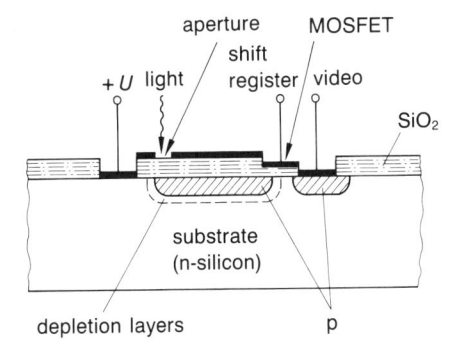

Figure 4.34 Diagrammatic construction of a single cell from a self-reading MOS photodiode array [70].

Such cells can be arranged either in lines or two-dimensionally. In the line array (Figure 4.35) a shift register switches the p-n capacitors, which are loaded by a reverse bias, in sequence to a video line. The capacitors, which have been partly or fully discharged during an image cycle due to optical carrier generation, are recharged. The output signal from a diode line containing n elements has the form of an n-component pulse train. The amplitude of the pulses is proportional to the intensity of the illumination of the diode in question. When the diodes are distributed over a surface one video line is again sufficient. However, two shift registers are needed.

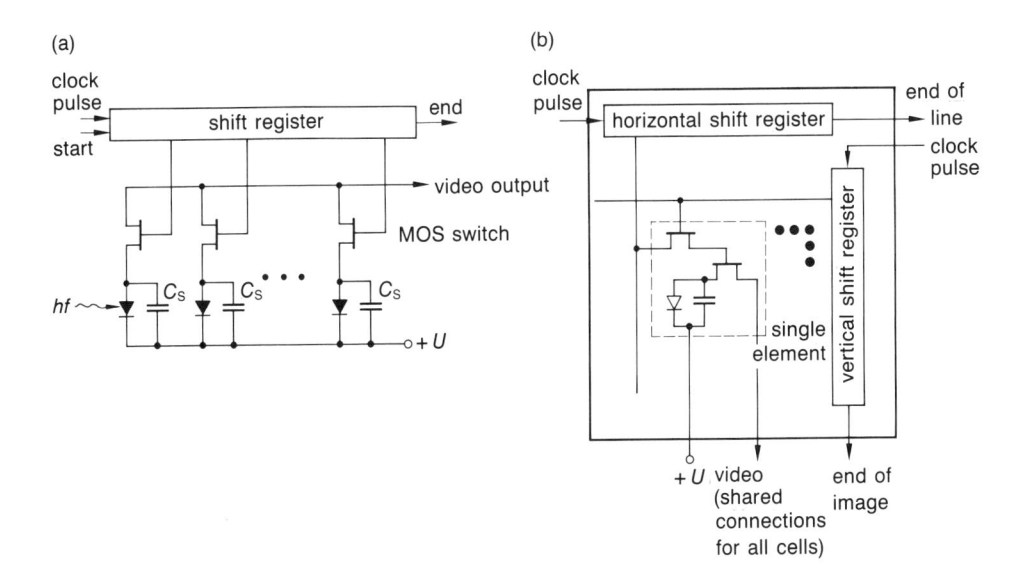

Figure 4.35 Equivalent-circuit diagram of (a) a one-dimensional and (b) a two-dimensional photodiode array.

44 Silicon sensors

Such photodiode arrays are commercially available as line arrays with between 64 and approximately 8000 image dots, or as area arrays with up to 256 × 256 elements. They are primarily used in automated process monitoring and are especially useful for the dynamic measurement of mechanical dimensions. Example applications are position recognition, the measurement of vibrations or oscillations, and character recognition.

4.6.4 Charge coupled devices

Charge coupled device (CCD) arrays have an important role to play as self-reading image sensors. The principle of charge coupling and the arrangements to which it leads are relatively simple. The minority carriers in an MOS structure are collected and stored in a localized potential well at a Si–SiO$_2$ junction. By applying the appropriate voltages at the metal electrodes it is possible to vary the potential in the semiconductor in such a way that the charges at a cell are shifted to the next one. A CCD is thus an analog signal shift register consisting of a row of closely packed MOS capacitors. The main feature of a CCD is that storage and transport are performed by separate elements with no depletion layer. In CCD image sensors the minority carriers are generated by the absorbed light during the integration period and are moved forward during each readout pulse until they appear as an image signal at the output diode in the form of a current pulse (Figure 4.36). Every third electrode possesses the same potential. If the three voltages, which differ in their value, are applied to the metallic contacts in the alternating sequence depicted in Figure 4.36, then the charge is transported to the right. Transfer losses occur and these are dependent on the shift frequency, the geometry and the number of cells to be processed. Here, drift and diffusion processes play a decisive role [77]. Charge losses may also occur as a result of surface states along the SiO$_2$–Si interface. These losses can be avoided by leading the carriers along a so-called buried channel. This is achieved by forming a layer of opposite conductivity and a thickness of

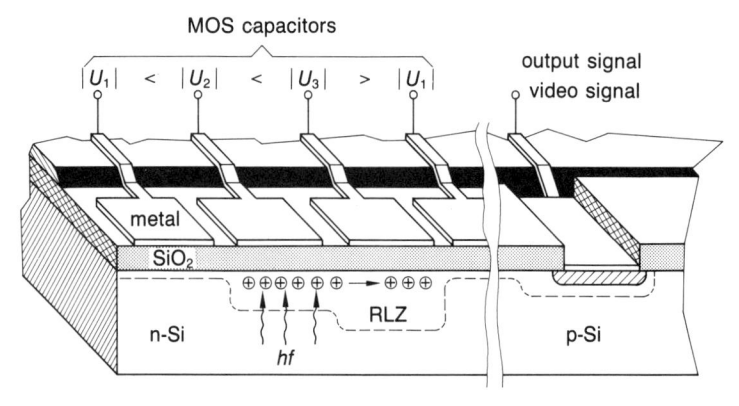

Figure 4.36 Diagrammatic structure of a CCD [70].

approximately 1 µm on the substrate. These CCDs are then known as BCCDs. Such BCCDs are more sensitive than silicon vidicons. CCDs can operate in either a three-phase configuration (three electrodes with different voltages U_1, U_2, U_3, as shown in Figure 4.36) or in a two-phase configuration ($U_0 \pm \Delta U$). Although the second alternative is simpler it requires a built-in asymmetry of the potential wells which can be obtained by varying the thickness of the oxide layer and the doping.

Light usually enters CCDs from the rear. The substrate must be thin (approximately 25 µm) if as many as possible of the carriers generated by the incident light are to be collected. This fact simultaneously determines the resolution of rear-illuminated CCDs since there is little point in using small electrode sizes.

The electrical output consists simply of a blocked p-n junction which converts the arriving charge packets into current pulses. In the case of a CCD line, this is an analog parallel–serial converter with timed integration of the optical input (Figure 4.37).

A CCD matrix (surface configuration) is read either directly (Figure 4.38(a)) or by a separate CCD memory (Figure 4.38(b)). In the first instance the stored image is transferred to the horizontal output register via pulse A. The output register is polled using the much faster pulse B and supplies exactly one image line before the next horizontal line is stored by means of pulse A. In the second case the entire image recorded in a memory line is read line by line into a non-photosensitive CCD memory from which it is then read as in the directly read variant. The advantage of the second arrangement lies in the fact that the image integration and readout processes are separated. This protects against blurring of the image.

If photodiode arrays and CCDs are compared, it can be seen that they both possess much the same sensitivity. Within the range 450–900 nm the sensitivity is also sufficiently uniform to make colour recording possible. Given optimal fabrication and operating modes, the dynamic range of a CCD can exceed that of a photodiode. CCDs are primarily used as an alternative to cathode ray tubes. Although they are still inferior to traditional television camera tubes (vidicon), many industrial applications do not require the resolution which these provide. Their small size and low power consumption gives CCDs important advantages over vidicon tubes. They are used for high-performance character recognition systems, contour recognition, parts identification by robots, image acquisition and comparison, evaluations of surface quality and other applications. CCD lines are found in fax

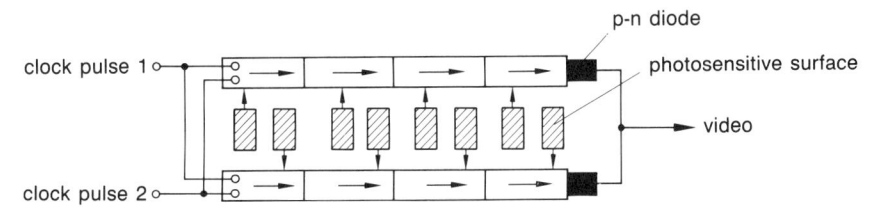

Figure 4.37 Readout procedure from a linear CCD.

46 Silicon sensors

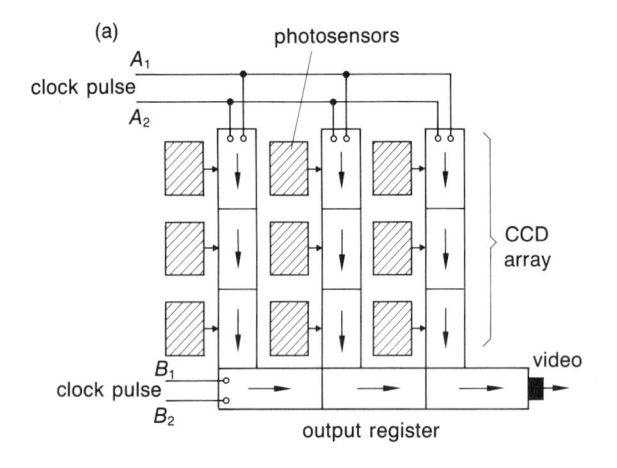

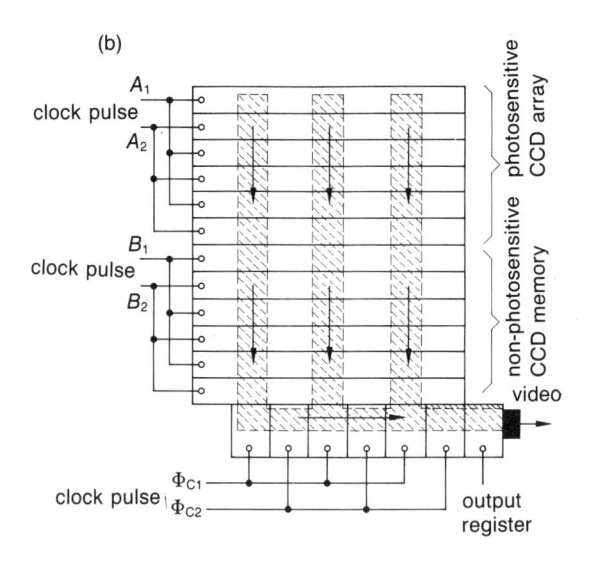

Figure 4.38 Readout from a two-dimensional CCD [70]: (a) direct line-by-line readout via a CCD output register; (b) column-by-column transfer to a non-photosensitive buffer.

systems. Nowadays line sensors with 1024 to 2048 image dots and matrix systems with 756×384 image elements are manufactured as standard. However, the development laboratories have already achieved arrays of 1000×1000 image dots. Type TH 7862/8400 is an example for a matrix sensor manufactured by Thomson-CSF.

Future developments aim to improve the resolution, to optimize the performance data and to produce sensors with colour capability. Solid-state image sensors with a photoconductive layer represent an interesting avenue of development

[78]. In such sensors an additional photoconductive layer system, for example a-Si:H, is applied to the single-crystal substrate. The construction of this type of image dot cell is presented in Figure 4.39. Image conversion is performed in the wide-area photoconductive layer system through the separation of the optically generated electron–hole pairs in the applied electric field. The substrate side of the photoconductive layer is connected with the source and drain of the selection transistors of the evaluation circuit via a matrix of metal contacts. The image information is read via the open selection transistors as in monolithic image sensors. Compared with diode or CCD arrays, such sensors which make use of a-Si:H possess considerable advantages. They possess the widest spectral sensitivity and their dynamic behaviour is excellent. However, their development is only just beginning.

4.6.5 Other semiconductor materials for optical sensors

Thanks to the high level of development in silicon technology, silicon-based optical sensors have established themselves in industrial applications. Despite this we should make a comparison with other materials at this point in order to point out the use of optical sensors in other wavelength ranges. Table 4.3 lists the materials which are commonly used today in applications in the visible and infrared ranges. The spectral sensitivity of silicon peaks at approximately 850 nm. This means that light sources such as lasers or luminescence diodes with class III-V connections (GaAs, GaP, mixed crystals) can be easily adapted for use with silicon detectors. This is a great advantage. Silicon is not very well suited for use with optical fibers since the ideal wavelength for optical communications at which the least attenuation and signal dispersion occurs is approximately 1.55 µm. Researchers are attempting to develop the ideal detector for this wavelength using the heterogeneous semiconductors

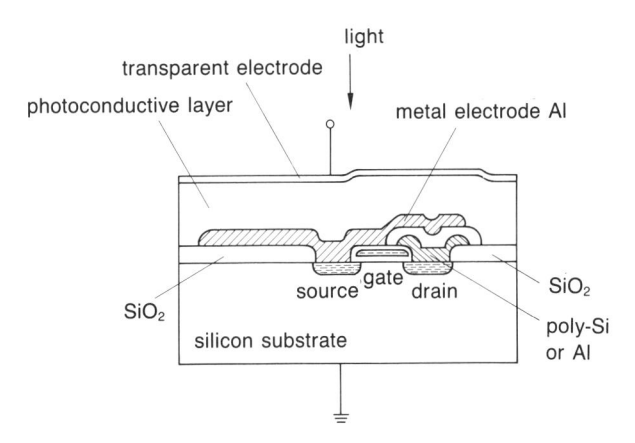

Figure 4.39 Diagrammatic construction of an image dot cell in a solid-state image sensor with photoconductive layer.

Table 4.3 Sensor materials and sensitivity range

Material	Range of spectral sensitivity (µm) at an operating temperature $T = 20°C$
Si	0.4–1.1
Ge	0.5–1.8
GaAs	0.7–0.9
InGaAs	0.8–1.8
InAs	1.0–5.5
InSb	1.0–5.5 (operating temperature 77 K)
PbS	1.0–3.5
PbSe	1.0–6.0

AlGaSb, GaInAsP or HgCdTe. Layers of these materials are deposited on monocrystals with a somewhat simpler composition, for example HgCdTe on CdTe. Quaternary compounds not only allow greater flexibility but also make possible the ideal wavelength setting relative to the fiber and the sensor.

Nowadays silicon sensors and GaAs/GaP LEDs are used for the industrial application of fiber optics as a communication medium.

Germanium sensors can be used in the infrared spectrum to 1.6 µm. InGaAs and InGaAsP photodiodes are gaining in importance. InSb detectors are employed for the mid-infrared range up to approximately 7 µm. However, cooling is necessary if optimum results are to be achieved. Infrared sensors are of great importance for temperature measurement and for thermal image processing. From the military perspective there is another significant 'wavelength window' at between 10 and 15 µm. $Hg_xCd_{1-x}Te$ compounds are being examined as possible detector materials at these wavelengths.

4.7 Magnetic field sensors

Magnetic field sensors (MFSs) are converters which are able to transform an existing magnetic field into an electrical signal. A distinction is drawn between two main groups of applications:

1. the direct use of an MFS as a magnetometer component [79], for example for measuring the earth's gravitational field, for reading magnetic codes or for monitoring magnetic equipment;
2. indirect use, i.e. the magnetic field simply functions as an information carrier for a non-magnetic signal, for example in contact-free switches, in the detection of changes in distance or angle, in zero potential current measurement or in an integrated watt meter.

Such a wide range of applications means that it is necessary to be able to detect magnetic fields down to the micro- and millitesla range. In recording media it is possible for stray fields of 10 µT to 10 mT to occur in the magnetic domain. Permanent magnets with fields of 5–100 mT are used in switches and in sensors for detecting changes of distance. Most MFSs employ the Lorentz force acting on charged particles moving in the magnetic field

$$F_L = e(\mathbf{v} \times \mathbf{B}) \tag{4.19}$$

where e represents the quantity of electron charge, $\mathbf{v}$ is the particle velocity and $\mathbf{B}$ is the magnetic induction. Starting with the relationship $\mathbf{B} = \mu\mu_0\mathbf{H}$, where μ is the magnetic permeability of the sensor material, it is easy to distinguish between two main groups of MFSs.

The first are MFSs which employ materials with a high permeability (ferromagnetic and ferrimagnetic materials). When $\mu \gg 1$ sensitivity is increased. Examples of this type of MFS are those based on NiFe thin films, the magnetostriction of nickel-clad fiber optics or magneto-optic effects. These sensors will be presented in greater detail in later chapters.

The second group are MFSs which make use of low-permeability materials (diamagnetic or paramagnetic materials). In such materials $\mu \sim 1$ is of no particular use. These materials comprise all those that use galvanomagnetic effects in semiconductors.

The following exposition will discuss this second category. The term 'galvanomagnetic effect' is understood to refer to effects such as Hall voltage, magnetoresistance and magnetoconcentration [80].

4.7.1 Galvanomagnetic effects

Figure 4.40 illustrates the Hall effect and the change in the resistance of the magnetic field. Two voltages can be measured on a semiconductor strip. If the semiconductor is homogeneous and located in the field-free space the dashed lines of flux are obtained (Figure 4.40(a)). The Hall voltage is zero since electrodes 3 and 4 share the same potential. If the magnetic field now acts vertically on the semiconductor layer the carriers are deflected from the original direction of the supply current I_0 by the Lorentz force and charge the lateral sides either positively or negatively. An opposing electrical field, the so-called Hall field, is generated. Charging ends when this field is exactly strong enough to counter the effect of the magnetic force on the electrons.

The equipotential lines are rotated through the angle θ_H (Figure 4.40(b)). A Hall voltage is present at 3 and 4. When the pick-up resistance value is high, the Hall voltage is generated by the positive or negative space charges along the sides of the semiconductor strip. At the same time a higher voltage U_R is measured since the resistance increases with the magnetic field, irrespective of its sign. The current paths are also depicted in Figures 4.40(b) and 4.40(c). These are rotated through the angle θ_H at the supply electrodes. Because of their high level of electrical conductivity the electrodes always represent equipotential lines. At the same time the lines of current

50 Silicon sensors

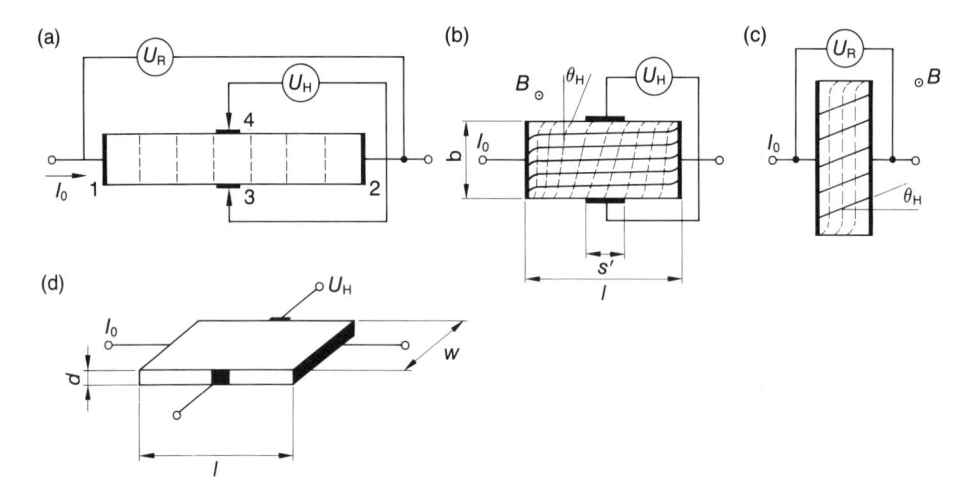

Figure 4.40 (a) Configuration for the measurement of the Hall effect (U_H) and resistance (U_R) in a semiconductor; (b) influence of the magnetic field on the distribution of current paths (−) and equipotential lines in a Hall geometry, and (c) of a plate geometry (B perpendicular to plane of diagram); (d) geometrical representation of the semiconductor plate.

must be parallel to the lateral edges since at equilibrium no carriers can leave the semiconductor. Rotation of the current paths in the vicinity of the control current electrodes leads to an increase of resistance in the magnetic field which can be measured with U_R. When the Hall voltage is U_H, then:

$$U_H = (R_H/d)I_0 B \tag{4.20}$$

The Hall voltage is proportional to the current I_0, the magnetic induction B and the quotient obtained from the Hall coefficient R_H and the thickness d. This means that thin layers with a high resistivity are required if a high level of sensitivity is to be obtained. Sensors which function according to this principle are known as *Hall generators*. The Hall coefficient at negligible minority carrier concentrations is given by:

$$R_H = -r/en \quad \text{for n-SI}$$
$$R_H = r/ep \quad \text{for p-Si} \tag{4.21}$$

The factor r allows for scattering effects in the real crystal. It can assume a value between 1 and 2 (for n-Si with a low doping level it is approximately 1.15). n and p are the carrier concentrations. To obtain finite dimensions in the semiconductor layer (length and width are comparable) it is necessary to introduce a geometrical correction factor into equation (4.20). Thus

$$U_H = (R_H/d)I_0 B G(l/w) \tag{4.22}$$

For example, for $2 \leqslant l/w \leqslant 3$, the value of G is 0.95–0.99.

The factor G is of interest since completely different Hall geometries can have the same G factor (Figure 4.41). This means that from the technical point of view it is not easy to manufacture rectangular laminae where $s/l < 0.1$ (Figure 4.41(a)). However, the same geometrical factor can be obtained using a cross-shaped configuration which is much easier to produce.

The Hall angle θ_H is characterized by the following relationship:

$$\tan \theta_H = \mu B \tag{4.23}$$

where μ is the Hall mobility of the electrons. Since $\tan \theta_H$ is also exactly proportional to the ratio of the Hall voltage U_H to the reference voltage U_R

$$\tan \theta_H \sim U_H/U_R \tag{4.24}$$

it follows that

$$U_H \sim U_R \mu B \tag{4.25}$$

This means that electron mobility is crucial for the detection of magnetic fields. A sensitive Hall probe should therefore possess a high degree of electron mobility. Another requirement is that the Hall coefficient is not too small since internal resistances of 100–1000 Ω, such as are usually found in current sources and amplifier circuits, have to be obtained.

To use the magnetoresistive effect it is necessary to measure the change in voltage $U_R(B)$, where

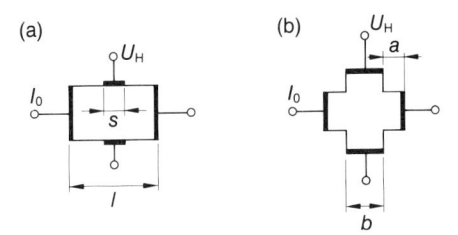

$$U_R(B)/U_R(0) = 1 + f_R(B) \tag{4.26}$$

Figure 4.41 Different geometries for Hall plates.

52 Silicon sensors

The function $f_R(B)$ is independent of the sign of the induction (B) and when the value of B is small it is proportional to B^2 (Figure 4.42). If the gap between the control electrodes is reduced, then the proportion of the change in resistance in the vicinity of the electrodes increases because of the proximity of the rotation of the current paths. This means that, given a uniform magnetic field, the resistance of the semiconductor layer increases as the ratio l/w decreases. In order to obtain semiconductor strips with a basic resistance of a few hundred ohms, rectangles of this type are connected in series. This is done by attaching parallel short-circuiting strips which run perpendicularly to the direction of current to a semiconductor strip (Figure 4.43(a)). The magnetic field causes the current paths to rotate by an amount approaching the Hall angle θ_H. Technologically, it is sensible to realize these short-circuiting strips in the semiconductor by means of an InSb/NiSb eutectic bond. The InSb contains parallel pins made of NiSb with a specific resistance which is smaller than that of the InSb by more than 2 orders of magnitude. The pins then function as short-circuiting strips. Components consisting of this type of magnetic field-independent resistance are known as *magnetoresistors*. The magnetoresistive effect is of considerable importance for thin-film components with ferromagnetic layers. These will be discussed later.

Hall voltages and changes in resistance are complementary effects. The Hall voltage increases with the ratio l/w while the change in resistance decreases.

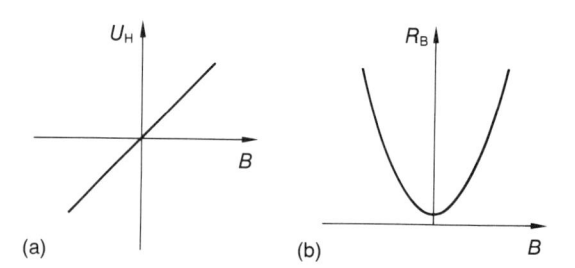

Figure 4.42 Dependence of (a) the Hall voltage U_H and (b) the resistance R_B on the magnetic induction.

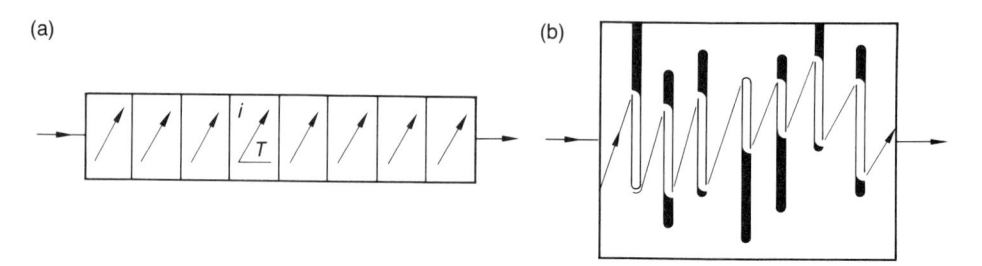

Figure 4.43 Current paths in the magnetic field (perpendicular to plane of diagram): (a) in a semiconductor layer with short-circuiting strips; (b) in an InSb/NiSb eutectic bond.

Magnetic field sensors 53

The magnetoconcentration effect is also included among the galvanomagnetic effects. This effect acts on the carrier concentration, making it dependent on the magnetic field [81]. This effect occurs when semiconductors have surfaces with different recombination speeds or when the volumes and surfaces of the semiconductors differ. This effect is important in so-called magnetic diodes.

4.7.2 Hall generators and magnetoresistors

Figure 4.44 presents a schematic diagram of a Hall generator. It is normal to use n-Si as the active material in Hall generators. In the illustration it takes the form of an epitaxial layer. Typical concentration levels have values of $n = 10^{15}–10^{16}$ cm^{-3}. The thickness of the active layer is 5–10 µm. This fulfils the requirements for a high Hall voltage in accordance with equation (4.20). The typical dimensions of a Hall plate are approximately $w \approx 200$ µm and $l \approx 200$–400 µm. The Hall generator, as depicted in Figure 4.44, is sensitive to magnetic fields which act perpendicularly to the chip surface. However, there are also special configurations which make it possible to measure magnetic fields which lie in the plane of the chip. Such arrangements are known as *vertical Hall plates* [84]. The great advantage of using silicon is that it provides full compatibility with IC technology. This means that the evaluation electronics can be integrated on the same substrate. Great use is made of this ability since the sensor signals emitted by the Hall generator are small (in the millivolt range) and have to be amplified. Figure 4.45 presents an example. The Hall layer is integrated in the base region of a differential amplifier [82, 83]. Commercial Hall generators have been supplied by many manufacturers (Honeywell, Siemens, Texas Instruments) for many years. Amplifier, power supply and evaluation electronics circuits are frequently integrated on the same chip as the

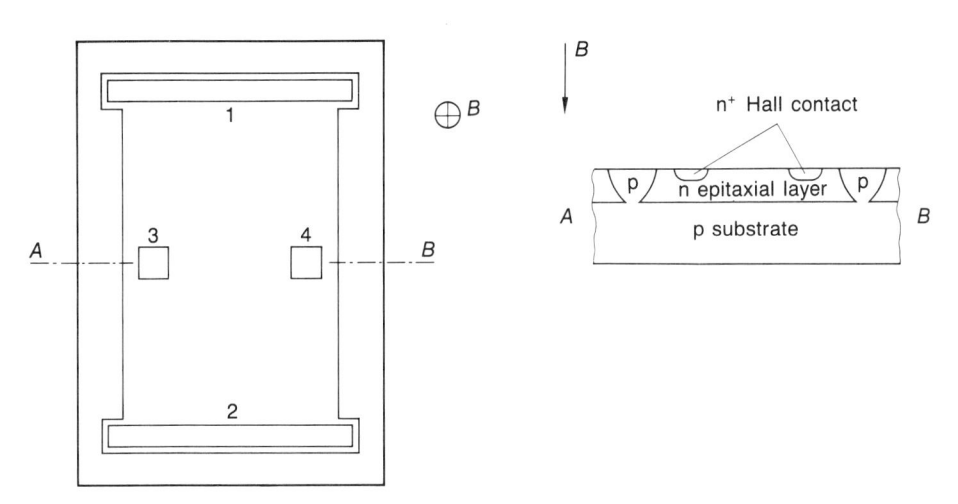

Figure 4.44 Hall generator, produced using bipolar methods, with current (1, 2) and sensor contacts (3, 4) [95].

54 Silicon sensors

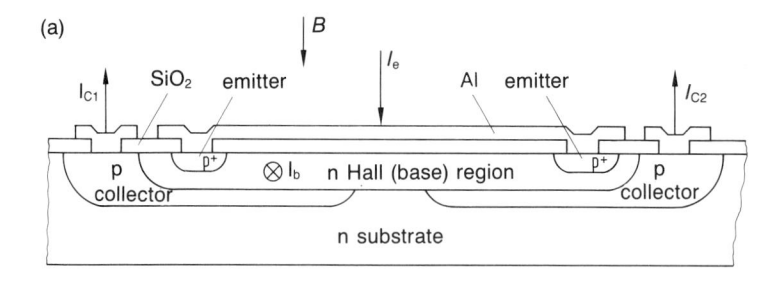

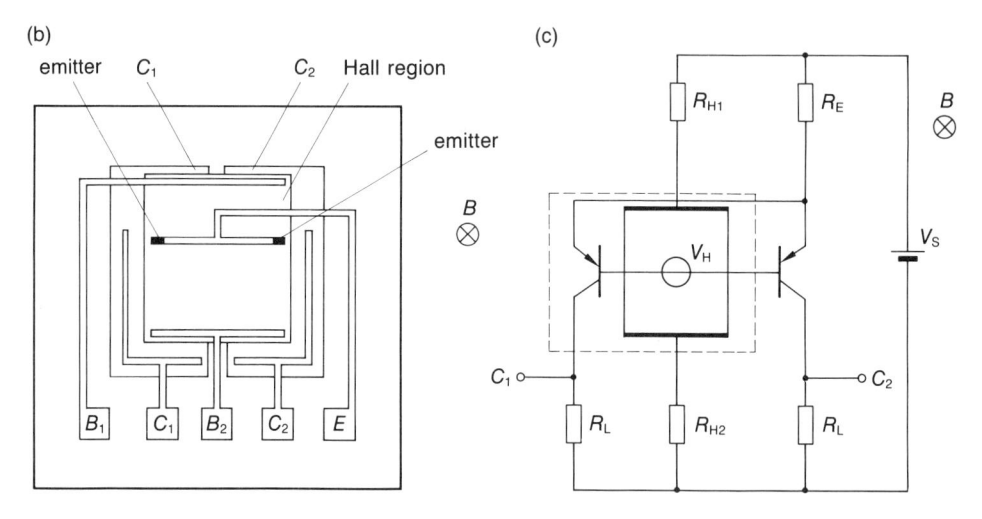

Figure 4.45 (a) Cross-section through a configuration which simultaneously uses the Hall layer as the shared base for a differential amplifier [95]; (b) top view of the configuration; (c) equivalent circuit diagram.

Hall generator. An analog output signal is obtained. If all that is required is a high–low signal, for example to provide a switch function, then digital Hall generators are used. Alongside on-chip amplification and voltage control, a Schmitt trigger and an output stage are integrated on the chip (Figure 4.46).

In fact, silicon is not very well suited for the use of the Hall effect since there are other materials with greater mobility (Table 4.4) such as InSb or InAs. However, due to the small bandgap these materials do not exhibit such low carrier concentrations at room temperature as silicon or GaAs. This unfortunately increases the overall loss of the Hall generator [80]. For this reason materials with a large bandgap and a high level of technical practicability should be selected. Other criteria relating to noise, offset and temperature dependence must also be met. This makes silicon and GaAs the most interesting materials. While silicon-based Hall generators can be used at temperatures of up to 200°C, those made from GaAs can operate at temperatures of up to 400°C.

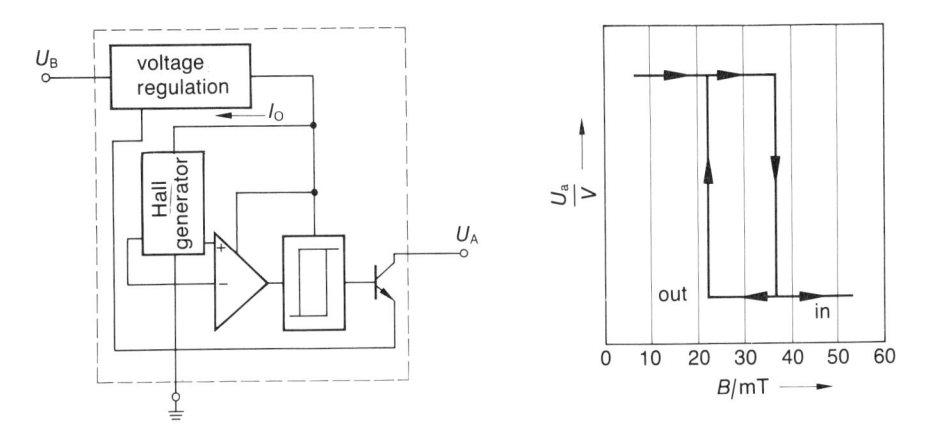

Figure 4.46 Principle of a digital Hall generator and characteristic curve.

Table 4.4 The electron mobility μ_n, bandgap E_g and intrinsic conductance concentration n_i at room temperature for various semiconductors

	InSb	InAs	Si	GaAs
μ_n (m² V⁻¹ S⁻¹)	7.7	3.0	0.15	0.8
E_g (eV)	0.24	0.45	1.12	1.43
n_i (m⁻³)	2×10^{22}	6×10^{20}	1.5×10^{16}	10^{13}

The use of superlattice semiconductors opens up extremely interesting avenues of development. The sensitivity achieved by these semiconductors is much greater than that obtained with the materials which have been used in the past.

Semiconductors with a high level of carrier mobility are required for magnetoresistive sensors (magnetoresistors) since the magnetic resistance effect is quadratically dependent on μ_n:

$$R(B)/R(0) = \rho(B)/\rho(0)\,(1 + \mu_n^2 B^2) \tag{4.27}$$

Classically, InSb has therefore been the preferred material. The magnetoresistors are constructed using eutectically bonded InSb/NiSb (see Figure 4.43). In order to achieve resistances of 100–1000 Ω the end of the semiconductor is ground to a thickness of approximately 20 μm and a meander path is etched out. The NiSb pins are aligned vertically to the direction of current and form short-circuiting strips. As it is necessary to compensate for the temperature dependence of the resistance of the magnetoresistor it is common for differential plate configurations to be used. These are magnetoresistors arranged in pairs at predefined intervals in which a locally varying magnetic field causes a difference in the resistance of the two magnetoresistors. This difference is evaluated as a voltage signal in a bridge

56 Silicon sensors

connection. Magnetoresistors are sensitive magnetic field sensors with a stronger output signal than is found in Hall generators. InSb is being increasingly replaced by ferromagnetic thin films.

4.7.3 FET-Hall sensors, magnetic diodes and transistors

Given certain modifications, a MOSFET structure can also be used as a Hall sensor [85–87]. Figure 4.47 shows the construction of such a sensor. The channel area between source and drain serves as the Hall element. The depth of this channel can be varied using the gate voltage and the drain–source voltage. As the Hall voltage U_H is directly proportional to the thickness of the Hall element layer it is possible to obtain voltages of up to 100 mV in a field of $B = 1$ T. The Hall voltage is tapped at two contacts located below the gate. Their position relative to the drain region is crucial if the FET is to exhibit optimum magnetic sensitivity. MOS-Hall sensors have already been used as integrated contact-free switches for keyboards [88].

The magnetoconcentration effect [89] can also be used in special structures (Figure 4.48). Holes and electrons are injected from the p^+ and n^+ areas into the

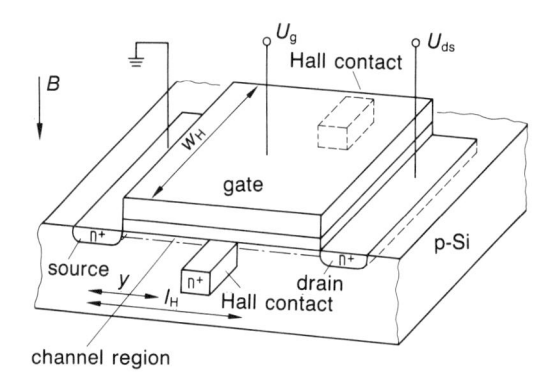

Figure 4.47 Construction of an n-channel MOS Hall element [95].

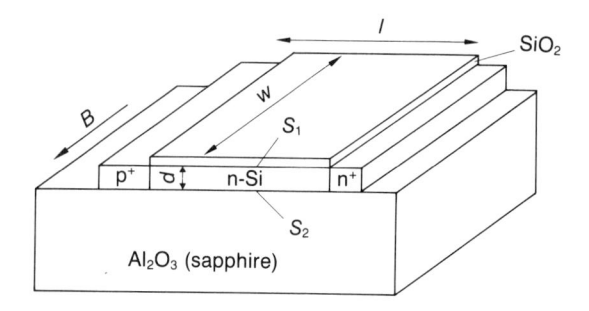

Figure 4.48 Silicon-on-sapphire (SOS) magnetic diode [95].

low doped n-area where they drift under the influence of the electric field. The recombination speed at the Si–SiO$_2$ junction is slower than at the base, the Si–Al$_2$O$_3$ junction at which S_2 is present. The magnetic field which is applied in the plane of the chip draws the carriers to one of the junctions, at which point there is a change in the current–voltage characteristic. If the thickness of the layer d is less than the ambipolar diffusion length and if the applied fields are less than 20 mT, then the variation of the diode voltage with the field, that is to say, the voltage sensitivity, is given by [90]

$$\partial U / \partial B = \frac{e(\mu_n + \mu_p)\tau_{\text{eff}}(S_2 - S_1)U^2}{8kTl} \qquad (4.28)$$

where l is the length of the n-area and τ_{eff} is the effective lifetime of the carriers. This equation is valid when the current flow is constant.

τ_{eff} is a complex function of S_1, S_2, d and the ambipolar diffusion length. The differing values for S_1 and S_2 can be generated by varying the roughness of the surfaces. In the case of magnetic diodes, sensitivities of 1–10 V/T can be obtained. A disadvantage of these sensors is the poor reproducibility of the surface with its high recombination speed.

In the case of magnetotransistors the output signal is provided by a differential current which is dependent on the magnetic field. In principle this type of sensor consists of a current source in the form of a p-n junction and a number of electrodes (collectors) which receive the current. A magnetic field acts on the currents which flow to the collectors. Magnetotransistors usually operate with two collectors. However, they are only sensitive to the presence of a B component. Two mechanisms contribute to the sensitivity of magnetic transistors, the deflection of the injected carriers by the Lorentz force and the modification of the emitter injection due to the creation of a Hall voltage in the base region [91–93]. These factors are differentiated by the direction of current flow relative to the chip surface and it is normal to speak of *lateral* and *vertical magnetotransistors*. In the lateral variant the current flows parallel to the surface of the chip and the sensor is sensitive to fields which are aligned vertically to the surface. Vertical magnetotransistors are sensitive to fields which lie in the plane of the chip. Current flow is perpendicular to this plane. Figure 4.49 shows the top view of a lateral p-n-p transistor which is sensitive to magnetic fields and which was first described in [94]. The emitter current which is injected into the base is deflected from its original direction by the hole–Hall angle θ_{Hp}. This results in a collector current difference I_C. If, in addition, a positive bias voltage is applied between the base contacts B_1 and B_2 the base region acts as a Hall plate at whose sides charge accumulates. The generated Hall angle field also changes the current path by an additional Hall angle θ_H. The current differential is

$$\Delta I_C = k(\mu_{\text{Hp}} + \mu_{\text{Hn}})BI_C \qquad (4.29)$$

58 Silicon sensors

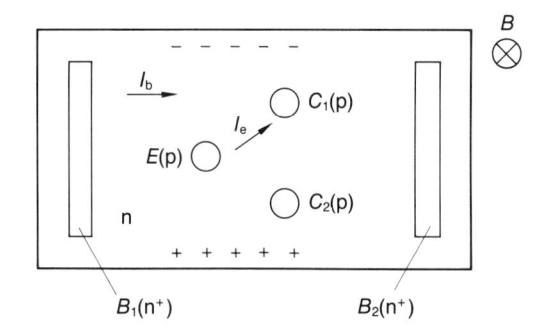

Figure 4.49 Top view of a lateral p-n-p magnetic transistor.

The structure of a lateral magnetotransistor is presented in Figure 4.50. Such a device exhibits both carrier deflection and an asymmetrical carrier injection which results from the Hall voltage which appears around the emitter in the base region [91].

A major aim is the development of a magnetotransistor which can simultaneously determine all the components of a *B* field. This is known as a *three-dimensional magnetosensor* and was first mentioned in [95]. Such a sensor is composed of a vertical two-dimensional magnetotransistor with a lateral one-dimensional magnetotransistor contained on a single chip. Its structure is illustrated in Figure 4.51. The preferred application range of magnetotransistors is for magnetic flux densities of $10^{-5} \leqslant B/T \leqslant 10$. This type of transistor can be used in temperatures between $-40°C$ and $150°C$.

The so-called *carrier domain magnetic field sensors* are very different from those described above. In these sensors the current is not distributed over the entire structure but is concentrated in fine current paths, also known as carrier domains. The first research work to investigate these sensors appeared in 1975 [96]. In most cases these sensors have p-n-p-n or n-p-n-p structures with complex current flow

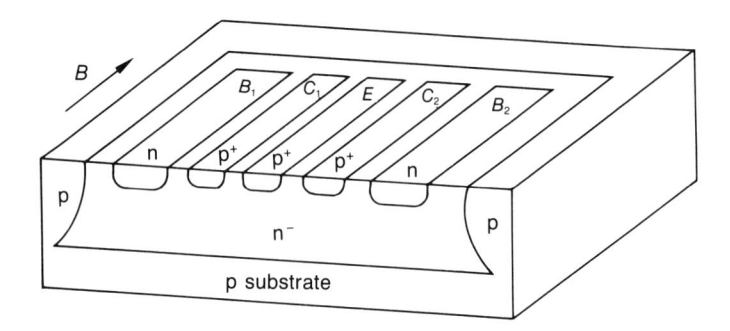

Figure 4.50 Structure of a lateral magnetic transistor.

Magnetic field sensors 59

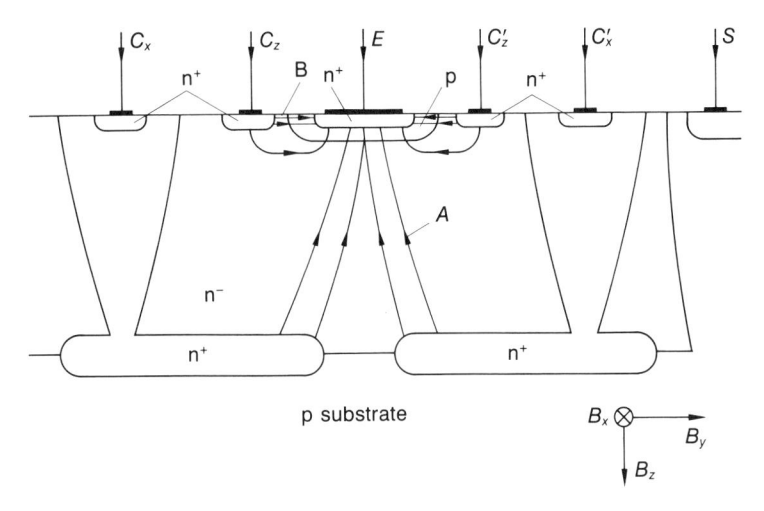

Figure 4.51 Cross section of a three dimensional magnetic field sensor. The vertical component (A) of the collector current is sensitive to B_x and B_y. The lateral component (B) of the current is used to detect B_z [87].

mechanisms. Vertical structures are also available individually [97]. The structure of this type of sensor is depicted in Figure 4.52 [98–100]. If a magnetic field is applied perpendicularly to the current paths, then the Lorentz force deflects them from their original orientation. When circular structures (combining n-p-n and p-n-p structures) are involved, the carrier domain starts to rotate. The frequency of rotation is a measure of the field B

$$f_r = d\mu_p B_z / (2\pi t_p R) \qquad (4.30)$$

where d is the radial distance between the n-p-n emitter region and the p-n-p base region, $\mu_p B_z$ is the Hall angle, t_p is the emitter–collector delay time and R is the external radius of the border of the n-p-n emitter. Equation (4.30) only applies

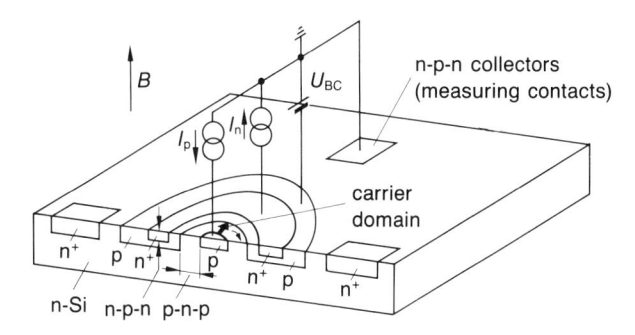

Figure 4.52 Structure of a rotating carrier-domain magnetic sensor [95].

beyond a threshold value of B. This magnetic field sensor is presented here because it outputs a frequency-analog signal.

4.7.4 Possible applications of magnetic field sensors

Because of the great variety of possible uses of Hall sensors and magnetoresistors it is only possible to present a few selected examples here. For reasons of cost Hall sensors used to be employed individually, whereas magnetoresistors are used in a differential configuration to compensate for the temperature dependence of the resistors. However, differential configurations of Hall sensors (for example, magnetotransistors or magnetic diodes) have been constructed. These contain, for example, four Hall generators on one chip [101] and constitute high-sensitivity Hall sensors.

Hall sensors can be used as analog sensors with a linear output signal which changes with the magnetic field, or they can be employed as Hall switches. The type chosen depends on the task to be performed. All galvanomagnetic sensors operate practically without wear and in consequence have an almost unlimited lifespan. In contrast to capacitive and inductive position sensors, they provide a stable output signal form which is not delayed and they are able to detect movements from 0–100 kHz.

An important application for Hall generators is the zero-potential measurement of electric currents (Figure 4.53). It is possible to measure direct and alternating currents of between 100 and 4000 A. Soft magnetic yokes can be used to increase sensitivity for the detection of smaller currents.

However, galvanomagnetic sensors are primarily employed as displacement and position sensors and as measurement instruments for speeds and angles of rotation. It is their task to convert localized changes in magnetic fields into electrical signals for subsequent analog or digital processing. Figure 4.54(a) illustrates the relative vertical (d) and lateral (s) movement of a permanent magnet. Figure 4.54(b)

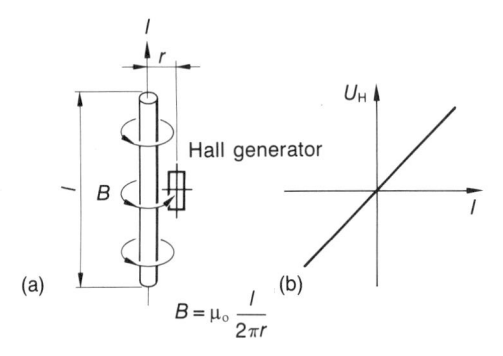

$$B = \mu_o \frac{I}{2\pi r}$$

Figure 4.53 Current measurement using an analog Hall generator (a: configuration, b: characteristic).

Magnetic field sensors 61

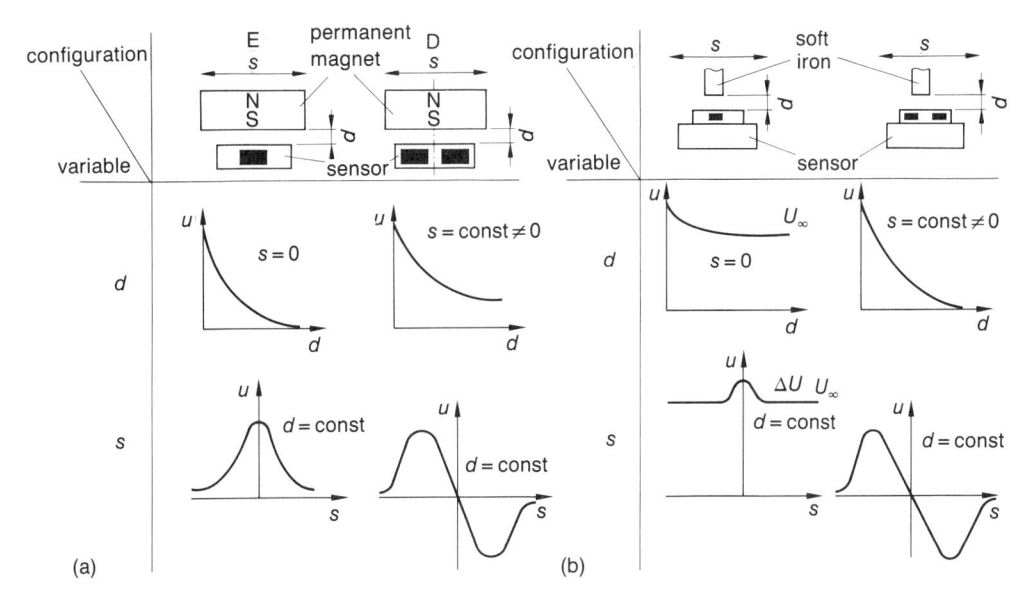

Figure 4.54 Basic configuration and diagrammatic characteristic curves for the detection of (a) the position of permanent magnets and (b) soft magnetic components using galvanomagnetic sensors (S: single, D: differential configuration) [4].

shows how the position of a soft magnetic object is detected by a so-called open magnetic circuit consisting of a permanent magnet and a sensor. U_∞ is the bias voltage supplied by the permanent magnet and ΔU is the actual measured voltage. Apart from these two main configurations, there are a large number of other possible arrangements. In the field of analog position detection, these sensors are thus used as analog distance indicators in which the linear components of the curves which are antisymmetric to $s = 0$ are used. This arrangement can also be employed in the conversion of mechanical vibrations into electrical oscillations, in pressure detectors where the sensors are combined with pressure cells (this requires a membrane connected to a fixed magnet), or for measuring power (using a magnet on an elastic spring), torque (torsion rod with an attached ring magnet) or acceleration (spring-bulk system in which the magnet functions as the bulk). The curves symmetrical to $s = 0$ can be used to detect final positions. This is useful in the line length control of printers or typewriters as well as for determining the stop position in transport devices.

Magnetoresistors are used in contact-free potentiometers [102]. Galvanomagnetic sensors are suitable for retrieving information which has been coded using soft iron components or permanent magnets located in a ferromagnetic material. An interesting application for analog Hall sensors is the collector-free direct current motor (Figure 4.55). When the control current through the Hall sensor is constant, the level and polarity of the Hall voltage indicate the position of the rotor

62 Silicon sensors

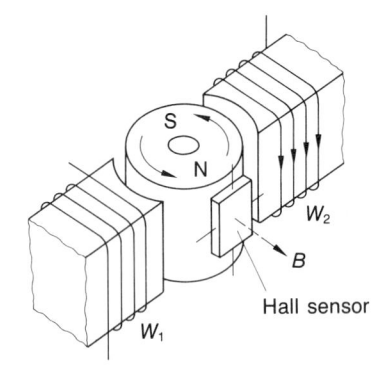

Figure 4.55 Principle of a collector-free direct current motor. The Hall generator registers the position of the permanent magnet relative to the fixed coil.

irrespective of the speed of rotation. In this way the current in the stator windings can be electronically regulated. The advantages of this type of motor are the excellent regulation of the speed of rotation, low operating noise emission and freedom from electromagnetic interference. Example applications are in video recorder heads or in floppy disk drives.

Analog and, to a greater extent, digital Hall sensors have many applications in the field of non-contact measurement of the speed and angle of rotation. Figure 4.56 illustrates a Hall sensor which uses a radial ring magnet to measure any angle of rotation at a high degree of accuracy of up to 0.2°. Speeds of rotation and angles can also be measured directly at rotating gear wheels without the need for a ring magnet. This is achieved by placing a magnet behind the Hall sensor. The flux density flowing to the sensor varies with the alternation of cogs and air gaps. If the Hall sensor is opposed to a cog made of a ferromagnetic material, then the magnetic flux density increases. The variations in flux density caused by the rotating gear wheel are detected by the sensor and are converted into square pulses. Digital Hall sensors are also used in the positioning of cylinders, cams, levers and shafts as well as for fuel-saving ignition timing in cars. In this type of application the fact that these devices are not damaged by dirt or moisture is particularly important. Another major application is their use in non-contact keyboards, for example in computers.

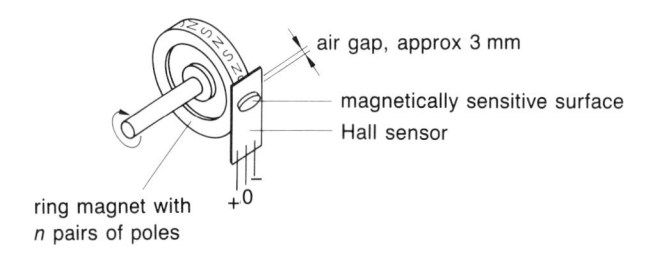

Figure 4.56 Hall sensor for measuring angles of rotation.

4.8 Micromechanical sensors

Pressure sensors with thin silicon membranes containing diffused or thermally deposited stress-sensitive resistors have been established micromechanical products for nearly 20 years. However, the range of application of micromechanical technology increased greatly at the beginning of the 1980s. Currently sensor technology only represents one of the fields in which micromechanics can be employed. Compared to others, the most promising advantage of this technology is the ability to connect a miniaturized sensor and an actuator on the same substrate, if possible together with integrated electronics. Silicon is the material which makes it possible to solve many of the problems facing sensor technology. For this reason it also plays a decisive role in micromechanics. It is possible to produce sensors for pressure, power, acceleration, gas flow, radiation, sound, humidity and even chemical composition. Alongside silicon, other materials used include a variety of polymers (such as PMMA, PVDF), quartz, and ZnO films. As this entire area is currently undergoing considerable development, I shall only present selected examples here and point out a number of interesting developments.

4.8.1 Acceleration/vibration sensors

Using suitable etching and doping methods, it is possible to manufacture miniaturized components for the measurement of acceleration and vibration [106]. Figure 4.57 illustrates the principle underlying a simple acceleration sensor [107]. This sensor contains a bar with a thickness of 5–15 µm which is etched out of the

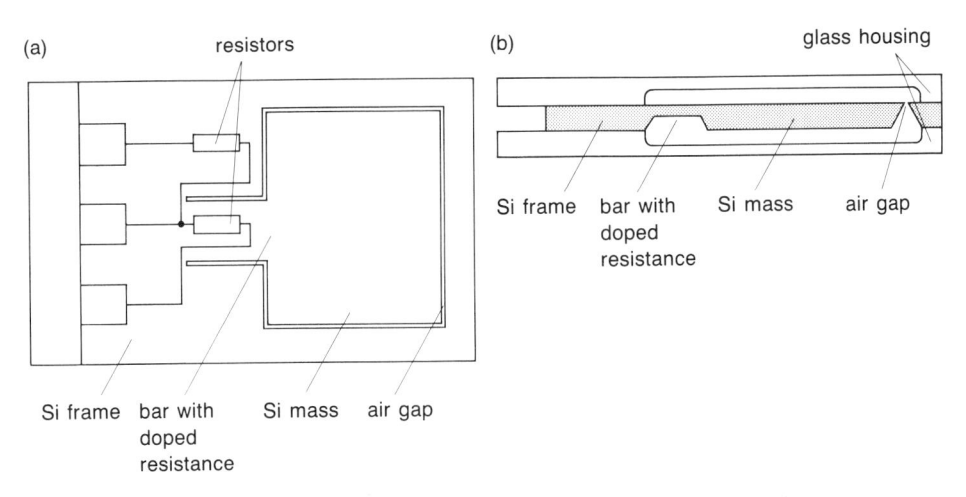

Figure 4.57 (a) Top view and (b) cross-section of a silicon acceleration sensor.

silicon substrate and which is fixed to it at one end. At the other end the bar supports a silicon lamella with a thickness of approximately 200 μm. Although this mass is supported by the bar, it is etched out of the substrate on all sides and can thus move freely. This mass is partly coated with gold or, in other versions, supports an additional mass. If the sensor is accelerated in a direction vertical to the surface, the bar bends like a rod which is supported on one side only. It is doped with resistors whose value changes as the bar bends. Capacitive, piezoelectric or optical means can be used to detect the deviation. When capacitive measurement is employed, the variation in the gap between the conductive reed and a metal plate positioned above it is determined. When optics are used, glass fibers register the deflection. The sensitivity and bandwidth of this type of sensor can be varied across a wide range by selecting the appropriate geometry. The sensor illustrated in Figure 4.57 is fused into glass, weighs approximately 0.02 g and measures 2 mm × 3 mm × 0.6 mm. Depending on its geometry, its measuring range extends from 10^{-1} to 10^3 m s^{-2}. The frequency of resonance is of the order of several hundred hertz. This principle has since been modified and the geometry layer structure and signal detection have been varied to obtain the sensitivities and measuring ranges required by particular applications. Figure 4.58 depicts the structure of a standard acceleration sensor produced by ICSensors [108]. In this sensor a freely vibrating mass has been separated from the silicon substrate. This is connected to the remaining frame by four arms. The deflection caused by the acceleration of the sensor bends the arms. Piezoelectric resistors are implanted at the points of greatest bending. These are connected to form a bridge circuit in such a way that, when the sensor is subjected to positive acceleration, two facing resistances increase, causing the other two to diminish. The structure of this sensor gives it a dynamic range which extends almost to a frequency of zero. This makes it possible to measure very low speeds and very low frequency vibrations. Standard sensors can measure accelerations of between $5g$ and $100g$ (g is the acceleration due to gravity). The frequency of resonance lies between 600 Hz (at $\pm 5g$) and 2750 Hz ($\pm 100g$). The acceleration sensors presented here are used in a variety of applications. They are an important component in ABS

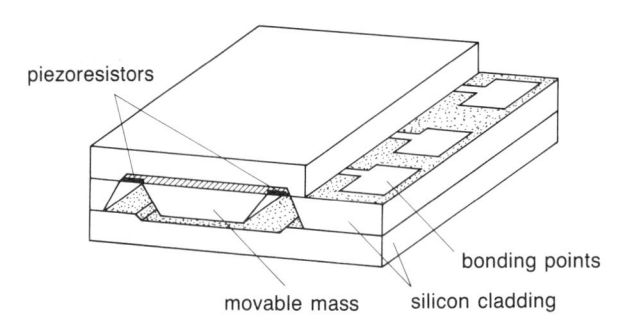

Figure 4.58 Three-layer construction of an acceleration sensor with automatic movement limitation [108].

systems and automobile safety systems, for example in the airbag system [109]. As high-sensitivity sensors ($10^{-6}g$!), they are also employed in space travel and in medical applications. Acceleration sensors are used in automatic washing machines, where they control the rinsing and spinning operations. As so-called g-switches, they are used in rockets and ammunition in military applications. In the future they will undoubtedly have an important role to play in the fields of machine and system diagnostics, robot control and navigational systems.

The technology has now been developed to etch not just one free-moving reed but a whole series of reeds which may be geometrically identical or different. Figure 4.59 depicts a representative of the second variant. With reed lengths of 0.2–1.7 mm and a thickness of 4 µm, the self-resonant frequency of these sensors is between 0.1 kHz and 10 kHz. [25] contains a description of a sensor with 50 reeds extending above a well etched out of silicon. Such sensors are very interesting for the measurement of sound, of sound conduction through solids and of vibrations as well as for diagnostic purposes.

4.8.2 Microbridge sensors

Figure 4.60 illustrates a sensor which is used to measure gas flow [111]. It is based on the principle of heat transfer. It is produced using the techniques of micromechanics and thin-film technology. It is remarkable for a number of reasons. The sensing element is composed of two suspended bridges (hence the title of this subsection) which are 400 µm wide, 500 µm long and span a channel of 130 µm which is etched in the silicon. The two bridges form a thermally insulated dielectric

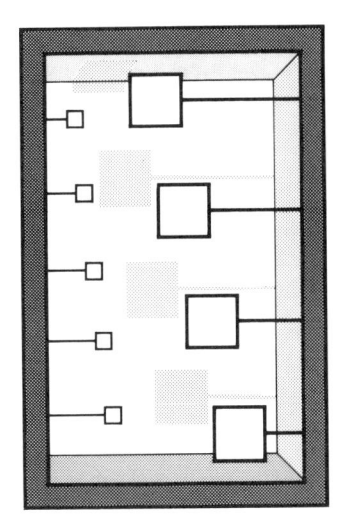

Figure 4.59 Vibration sensor with several gold covered silicon reeds with different bar lengths.

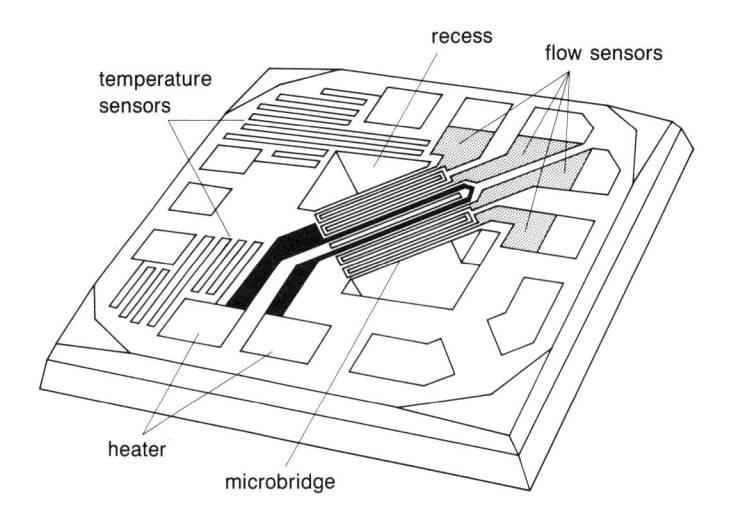

Figure 4.60 Silicon sensor for the measurement of gas flows.

layer. Both bridges contain thin-film resistors of several hundred ohms which are the arms of a Wheatstone bridge and which function as a sensitive layer. Another resistor which covers both bridges equally warms the measuring resistors. The difference between the heating temperature and the surrounding temperature is kept constant by means of a control circuit located on the chip. Air passes through the channel below the bridges. This air flow cools the measuring resistor on the front bridge and transfers its heat to the rear resistor on the second bridge. The output voltage of the Wheatstone bridge is a measure of the volume of air passing through the sensor. If the direction of air flow changes then the polarity of the voltage changes. For this sensor to function, it is necessary for the heating and measuring resistors to be thermally insulated from the rest of the chip. In addition, the dielectric layer must be so thin that it transmits no heat. In consequence all the heat is transferred by the air from the heater to the measuring resistors. The response time of the sensor is 3 ms. Air flow speeds as low as 1 cm^3/min can be measured. This sensor is very useful in medical equipment, especially for monitoring breathing, and in air conditioning and energy technology.

A so-called resonant microbridge has been suggested as an integrated sensor for detecting solvent vapours [112]. A thin polymer layer is applied to a poly-Si microbridge which is electrostatically made to resonate and which has dimensions of several hundred micrometres. This absorbs the solvent vapours and the resonant frequency changes as a result of the increase in the mass of the microbridge. A capacitive process is used to determine the value of this shift from the change in the vibration of the microbridge. Evaluation is performed using a monolithic integrated NMOS circuit.

It would be possible to list further examples of micromechanical sensors [114]. Many of them are still at the laboratory stage. Apart from silicon, quartz represents another interesting substrate material, especially in the case of power sensors [113]. Piezoelectric ZnO layers are important in both sensors and actuators. The most spectacular example so far of the potential applications of micromechanics is the prototype of a gas chromatographer on a silicon chip.

5 Sensors in thin-film technology

Many of the external influences which have to be recorded only act on a sensor's thin surface layer; light is an example of this. Alternatively, they may require a small bulk in order to achieve a fast response time, as in the case of heat. Certain functional aspects, such as the permeability to moisture of thin metal films, may make thin layers attractive. In addition, silicon, on which we have concentrated up to now, is only able to fulfil some of the functions required of sensors. Other materials, in the form of thin films, have to be used.

The key elements in a thin-film sensor are the substrate and the material of the thin film. The materials used for the substrate are glass, metal, plastics and, more recently, silicon. The use of silicon becomes interesting when the monolithic integration of the sensor and of the detection electronics is required. However, glass, ceramics and metals are most commonly used in substrates. Depending on requirements, it is possible to use either simple window glass or expensive quartz glass. Recently much interest has also been shown in crystalline Al_2O_3 (sapphire).

A number of materials can be used for the sensor films, from simple metal films and oxide layers to semiconductor layers. The following layer types are used:

- temperature-dependent resistance layers (e.g., Pt, Au, Ni, Ni alloys, platinel, a-Si:H, ZnO, PVF_2, $NaNO_3$);
- light-sensitive layers (e.g., CdS, PbSe, Si, a-Si:H, HgCdTe, organic polymers);
- pressure-sensitive resistance layers (e.g., NiCr alloys, poly-Si);
- piezoelectric layers (e.g., lithium niobate, ZnO);
- moisture-sensitive layers (e.g., Ta_2O_5, Al_2O_3, polystyrene, plasma polymerides);
- chemical-sensitive layers (e.g., SnO_2, ZnO_2, Fe_2O_3, ZrO_2);
- magnetoresistive layers (ferromagnetics, amorphous magnetics).

Thin films are typically between 0.01 and 100 µm thick. It is possible to specify an optimum thickness range for different classes of material, depending on the function that is required of them. The examples listed illustrate the varying manufacturing methods and the different areas of application. In the following a few examples from this wide range of possibilities will be discussed in greater detail.

Thin-film technology plays an important role in chemical sensors, the discussion of which will be left to Chapter 8. However, first it is necessary to make a few comments about the fabrication techniques.

5.1 Deposition techniques

A key part of what is to become the sensor is the substrate. In industrial production the largest possible substrates are used for the depositing of the layers, for example from 10×10 cm to 50×50 cm. These large surfaces are then structured and divided. Purity, structure, thickness and reproducibility are important aspects of the thin films which are deposited. In order to manufacture structured thin-film sensors, for example in a two-dimensional configuration, photolithography and, more rarely, masking techniques are used. The most important methods for the generation of thin films are

- thermal evaporation, electron beam evaporation;
- sputtering (direct current (DC), radio frequency (RF) and magnetron sputtering);
- ion-plating;
- chemical vapour deposition (CVD) in various forms.

In the vacuum evaporation process the material which is to be formed into the thin film is heated in a vacuum until it evaporates. Sputtering causes the material to evaporate by bombarding it with high-energy particles, Ar ions for example. Films produced in this way have considerably better adherence than in the case of vacuum evaporation. The growth rate can be accelerated by applying a magnetic field, for example. Ion-plating combines some of the advantages of the first two methods. The deposition material is evaporated using a classical method, for example an electron gun. At the same time an inert or reactive gas is injected into the reaction chamber and a plasma discharge is generated. This ionizes the evaporating material which is accelerated in a field towards the substrate. This process has beneficial effects on both the adherence and the morphology of the films. In the CVD process the desired material is deposited under the influence of gaseous materials which the temperature or gas discharges cause to decompose or react. Thus monosilane (SiH_4) decomposes to form poly-Si. This category of processes also includes plasma polymerization, in which thin polymer films are produced from the gas phase.

In principle every thin-film sensor is a multilayer device consisting of a number of individually structured layers. This is illustrated in the following examples.

5.2 Selected thin-film sensors

In some of the silicon sensors presented in the previous chapters the use of thin-film

70 Sensors in thin-film technology

techniques has been implicit. This is the case with modern optical sensors, the a-Si:H temperature sensor and the poly-Si pressure sensor. Chemically sensitive thin-films are very important, and will be discussed in detail later. In the following a number of different examples will be set out. A complementary overview can be found in [123].

The new generation of pressure gauges make use of fabrication technologies which involve integrated piezoresistive semiconductor resistors in monocrystalline silicon on the one hand, and technologies employing thin-film WSGs on metallic spring bodies on the other. Figure 5.1 illustrates a planar spring element shaped using a structure etching process [115]. The WSGs are located at the points of maximum strain and form a bridge structure. CuBe was chosen as the spring material. During production a 4 inches square CuBe spring plate substrate with a thickness of 0.5 mm is used. Sixteen spring elements can be manufactured on one 4 inch substrate. The thin-film system is structured first and this is followed by the structuring of the spring element. Figure 5.2 shows a cross-section through the thin-film system. This basically consists of an insulating layer, the WSG resistance layer and the contact layer. In addition, a passivation layer is inserted between the substrate and the insulating layer and a diffusion barrier is inserted between the resistance and contact layers. After coating, lithographic processes are employed to structure the layers. There are a number of reasons for the complexity of this layer arrangement. For example, the task of the Ta layer is to prevent both underetching of the SiO_2 during spring structuring and corrosion of the CuBe at high temperatures. Considerable demands are placed on the WSG resistance layer, which should be considered as part of a system which includes the supporting layer and the contact layer and which requires optimization.

When suitably encapsulated, this type of spring element–thin film–WSG system possesses a number of useful characteristics. It can be used over a wide range of temperatures, the creep of the WSG is low, the bridge resistance is high and the system can be manufactured at low cost.

Using CVD technology, it is possible to apply poly-Si to a spring body. The

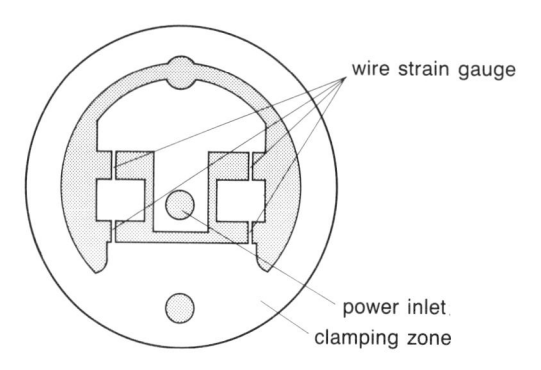

Figure 5.1 Spring element of a pressure sensor.

Selected thin-film sensors 71

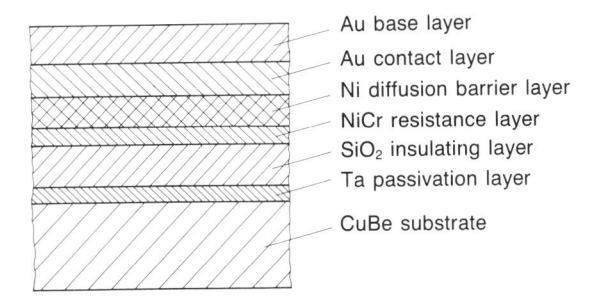

Figure 5.2 Cross-section through a thin-film system.

Au base layer
Au contact layer
Ni diffusion barrier layer
NiCr resistance layer
SiO$_2$ insulating layer
Ta passivation layer
CuBe substrate

support can be made of the monocrystalline silicon which was mentioned earlier (Section 4.5.2) or it can equally well be made of a metal such as CuBe or triple layers of CuBe/AgPd/CuBe [116]. This type of silicon pressure sensor has both a strong output signal and a low temperature coefficient. It is to be expected that poly-Si WSGs will plug the gap between the low-cost sensors with WSGs on monocrystalline silicon and the precision sensors which use metal WSGs.

Recently, sensors in which the strain-sensitive material is an a-Si:H film formed on a thin metal membrane have been presented [117].

Traditional temperature sensors, such as thermocouples, thermopiles or metal resistance thermometers, are made of wire. However, thin-film technology is now also encroaching on their territory. Thin-film resistance thermometers possess a two-dimensional, structured metal film made of Pt, NiCr or other metals formed on an isolating glass or ceramic substrate. The resistance of the metal possesses a fixed temperature dependence. Film sensors are smaller, more robust and possess a better response time than wire thermometers. Their disadvantage is that they possess a low temperature limit which results from the emergence of diffusion phenomena at over 550°C.

Thermocouples and thermopiles can also be constructed from thin films (see Figure 4.10). A thin-film thermocouple for measuring radiation [118] is depicted in Figure 5.3. This type of configuration can be used to advantage if it can be fixed to a flat measuring surface. Even high temperature sensors, produced for use in aircraft turbine blades, are developed using thin-film sensors [118]. However, this

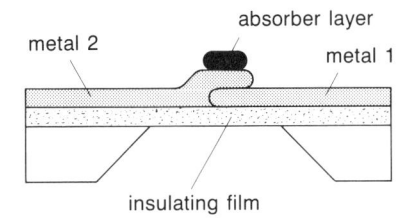

Figure 5.3 Cross-section through a thin-film thermocouple for measuring radiation.

72 Sensors in thin-film technology

requires the use of special substrates and the implementation of complex sequences of layers.

Magnetoresistive sensors represent an important group of thin-film sensors. These make use of the so-called *magnetoresistive effect* which was discovered by Thomson in 1857 in the course of an examination of NiFe alloys. However, it was only with the development of microelectronics and the accompanying thin-film techniques that this effect aroused widespread interest. Magnetoresistive sensors are produced by applying ferromagnetic materials (NiFe, NiFeCo alloys) with thicknesses of 20–100 nm to silicon, glass, ceramic or ferrite carriers by means of evaporation or sputtering techniques [120]. During coating, a magnetic field is applied parallel to the surface of the carrier. This results in a predetermined, uniform orientation of the preferred magnetic axis in the layer. Magnetizing the ferromagnetic layers gives these sensors properties which differ from those induced by the Hall and magnetoresistive effects in semiconductors or non-magnetic metals and invests the devices with a number of advantages.

To explain the way this effect operates, consider a magnetoresistive strip as depicted in Figure 5.4(a). The maximum resistance value R_o of the sensor strip occurs when there is no external field. If a measuring field is applied in the y-direction, the magnetization is rotated from the x-axis and the resistance falls. The lowest resistance occurs when the magnetization is rotated through 90°, that is to say, when it is aligned in the y-direction. This requires a magnetic field H_y with a minimum field strength of

$$H_y = H_k + (d/w)M_S \qquad (5.1)$$

where H_k is the so-called anisotropic field strength which characterizes the magnetic

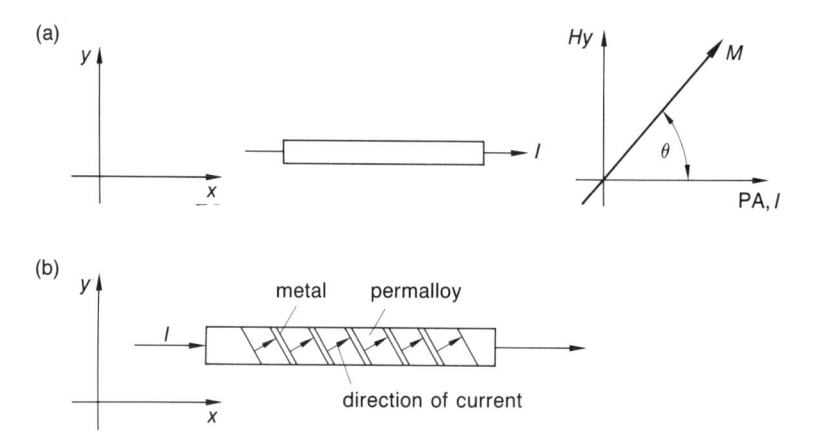

Figure 5.4 (a) Magnetoresistive strip, where the direction of the measured current is parallel to the preferred axis PA; (b) sensor strip with barber's pole configuration [120].

properties of the material, d is the depth and w the width of the sensor strip and M_S is the saturation magnetization. If the dimensions d and w are selected correctly, and given that l (length of strip) is much greater than w, then it is possible to adjust the minimum field strength necessary for a 90° rotation within wide boundaries.

If we let θ be the angle between the direction of magnetization which arises for a particular value of H_y and the direction of current, then the change in the resistance of the sensor is

$$R = \Delta R_{max} \sin^2 \theta \tag{5.2}$$

where $R_{max} = R_{\theta=90°} - R_{\theta=0°}$

At the same time the following applies to the angle φ between the direction of magnetization and the preferred axis:

$$\sin \varphi = H_y/H_0 \tag{5.3}$$

from which follows, assuming $\varphi = \theta$ (which is correct as long as the preferred axis coincides with the direction of current):

$$\Delta R/\Delta R_{max} = (H_y/H_0)^2 \tag{5.4}$$

Because of this quadratic relationship the transfer characteristic exhibits a high degree of nonlinearity, especially when the strength of the field H_y is weak. In some applications this is not a problem and may even be desirable. However, in general a linear characteristic is preferred and there are a number of ways to achieve this. One of the most common of these is the use of the so-called barber's pole configuration in which thin, highly conductive metal strips are fixed to the surface of the sensor strip at an angle of 45° (Figure 5.4(b)). The current through the high-

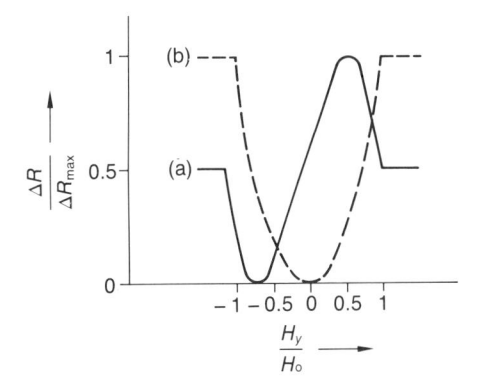

Figure 5.5 Curve for the function $\Delta R/\Delta R_{max} = f(H_y/H_0)$ (a) with and (b) without barber's pole configuration.

resistance ferromagnetic material then takes the shortest path between the metal strips, that is to say, at 45° to the preferred axis. Now $\theta = \varphi - 45°$ and equations (5.2) and (5.3) give

$$\Delta R/\Delta R_{max} = \tfrac{1}{2} + \Delta R_{max}(H_y/H_0)\sqrt{1 - (H_y/H_0)^2} \qquad (5.5)$$

Figure 5.5 presents the resulting curve. The central component is predominantly linear. Such barber's pole configurations are used in commercial magnetoresistive sensors [121]. In order to counter the high temperature coefficients of magnetoresistive materials, four sensor elements are connected to form a Wheatstone bridge. The resistance strip on each element has a meandering shape which gives it the greatest possible length. Such sensors are more sensitive than Hall sensors and can operate within a large temperature range. This increases the number of potential applications. They can be used to measure magnetic fields, to determine and monitor positions, as an instrument transformer in magnetic storage technology or for current and power measurements. An example of a commercial sensor of this type is the KMZ10B (Valvo) [122].

6 Thick-film sensors

Thick-film technology is a sensor fabrication technology which is microelectronics-compatible. So far it has been little used in sensor production, despite having an established position in the electronics world – somewhere between printed circuits with discrete components and integrated circuits. The reason for this neglect is the lack of practical experience in the production of sensor solutions using this technology and the absence of suitable paste systems. However, we can expect the importance of this technology to grow in the future since it can be used to produce medium-sized runs of sensors at low cost and permits the hybridization of sensor and evaluation electronics to produce 'intelligent' sensors. Thick-film technology can only be used to produce passive components such as printed circuits, resistors or capacitors. Active components have to be bonded to the substrate at a later stage, a requirement which is characteristic of hybrid technology. Sensor development requires suitable thick-film pastes with sufficiently sensitive properties. Table 6.1 lists the physicochemical effects associated with thick-film sensor pastes.

6.1 Production stages

Thick-film pastes are applied to a substrate using a screen printing process. The pastes are then dried at a temperature of 100–200°C and are burned in at temperatures of over 500°C [17]. During this process a number of different temperatures are used in sequence. Thin steel or nylon screens are used for printing. Stencils are placed over the screens and the glaze is printed through them. Multilayer systems can be produced by repeating the screen printing and temperature exposure process a number of times. The layers are typically 5–50 µm in thickness with conductor widths of 100–250 µm. Ceramics, enamelled steel, glass or flexible plastic substrates can be used as the substrate material. The most usual material is Al_2O_3. After burn-in, the substrates are snapped, broken or cut into the required shape.

The thick-film glaze usually consists of an active component (metal, metal oxide, ceramic), a flux (fritted glass, bismuth oxide) and organic solvents or bonding agents. The composition of the glazes can be manipulated to produce the required

76 Thick-film sensors

Table 6.1 Examples of the physicochemical effects associated with thick-film sensors

Resistance change ΔR

Linear resistance dependence of temperature (anemometer, throughflow)
Use of NTC or PTC effects (temperature sensor)
Change of conductivity between thick-film electrodes (concentration sensor)
Change of resistance of surface layers due to chemical reactions (chemical sensor)
Piezoresistive behaviour of thick films (pressure sensor)

Capacitance changes

Adsorption effects at surfaces (moisture sensor)

Thermoelectric voltage (temperature sensor)

Electrochemical potential (pH sensor, acidity sensor)

---

physical and technical properties such as toughness, coefficient of thermal expansion or the temperature coefficient of a resistor. Today conductive, covering, resistance and dielectric glazes are commercially available and these are used to produce the circuitry and the passive components as well as to protect the circuits. It is the resistance and dielectric glazes that are of particular interest in sensor applications. Alongside fritted materials and bonding agents, resistance glazes are predominantly composed of metal oxides such as ruthenium oxide and/or rhodium oxide. The character size of the thick-film resistance is influenced by processing parameters such as burn-in temperature, layer thickness, number of layer-forming procedures and others. Thick-film resistance sensors are manufactured with both positive (PTC) and negative (NTC) temperature coefficients. In the case of NTC resistors, the temperature coefficient can have a value between -500 and 7000 ppm/K. Dielectrical glazes usually consist of finely ground glass or ceramic materials with, for example, a high dielectric constant. Examples of such materials are $BaTiO_3$ and Ta_2O_5.

6.2 Examples

Table 6.2 lists the possibilities for realizing sensor elements in thick-film technology. It should be noted that if this wide range of possibilities is to be realized, then new glazes will have to be introduced or existing glazes modified. The production of such glazes requires a specific body of experience. Much work is currently in the development stage. The technology for the production of thick-film sensors for temperature measurement has been mastered [124, 125]. Glazes containing platinum or nickel are suitable for producing resistance temperature sensors. Many variants of the resistance glazes mentioned in Section 6.1 are employed in PTC and NTC temperature sensors. The field of pressure sensors is dominated by semiconductor sensors or film strain gauges.

Examples 77

Table 6.2 Sensors in thick-film technology

Measurand	Possible method of technical realization
Power, pressure	Thick-film, bridge-connected strain gauge resistors Capacitive thick-film pressure sensors
Temperature	Thick-film resistance thermometers (Pt, Ni) Temperature-dependent, dielectric thick-film materials NTC and PTC thick-film thermistors Thick-film thermocouples
Humidity	Capacitive humidity measurement using dielectric thick films Resistance hygrometer
Chemical measurands	Galvanic cells with strong electrolyte Thick-film lambda probe Ion-sensitive thick-film membrane
Position, path, angle	Thick-film potentiometer 2D position sensor
Flow, throughflow	Combination of thick-film filament resistor and temperature sensor

The piezoresistive behaviour of thick-film resistors which contain ruthenium can also be made use of. To this end bridge connections are formed on a substrate, for example Al_2O_3, which also serves as a membrane, and are used to measure pressure [126, 127]. Accuracies of 0.5% can be achieved. The electronic evaluation circuit can be realized using either thick-film or hybrid techniques. This is an important advantage. The principle of a capacitive differential pressure sensor in thick-film technology is illustrated in Figure 6.1 [128]. An important feature of this sensor is the use of Al_2O_3 membranes with electrodes applied using thick-film techniques. Hooke's law applies to this material up to breakdown, that is to say, it exhibits no hysteresis after a current has passed through it. This is a positive factor for long-term stability and reproducibility.

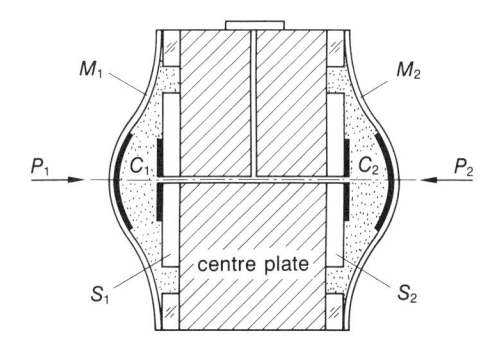

Figure 6.1 Differential pressure sensor using thick-film technology [128].

An important field for sensor technology, and one in which thick-film technology can lead to very interesting solutions, is that of chemical sensors. Solid electrolyte pastes such as ZrO_2 or metal oxide pastes based on SnO_2, ZnO, TiO_2 or WO_3 are used as the sensitive material. The heating elements which the sensors require can be produced using thick-film technology. This technology is also of interest in connection with moisture sensors which make use of changes in resistance, for example in $MnWO_4$ glazes [129] or in Al_2O_3 [130], or which function in accordance with capacitive principles [131].

The potential of thick-film technology for sensor development is as yet far from exhausted. Such sensors show immense promise for the future thanks to their high degree of reliability, their low space requirements and the ability to use hybrid techniques to produce medium-size runs of customer-specific circuits at low cost. Currently the most common areas of application are automobile and consumer goods technology, office technology, communications technology and military technology.

7 Fiber optic sensors

Low attenuation optical fibers, generally known as fiber optic cables (FOCs) were introduced in the mid-1970s and have revolutionized communications technology. The objective has always been to manufacture FOCs with transmission properties which are independent of external, changeable environmental parameters such as pressure, temperature and humidity. It is precisely these areas of sensitivity, which have been eliminated at such cost in optical fibers intended for communication purposes, that are desired in fiber optic sensors (FOSs). Intensive research into the development of FOSs began world-wide at the end of the 1970s [132, 133]. These sensors possess a number of crucial advantages over traditional electrical or mechanical sensors:

- suitability for a wide temperature range;
- resistance to corrosive environments and radioactive contamination;
- considerable protection against explosions;
- insensitivity to interference from electromagnetic radiation;
- miniaturization, leading to low production costs.

The dielectric properties of FOCs mean that they can operate at zero potential. They can also be used for data transmission where their low degree of signal attenuation and their high transmission capacities are exploited.

FOS production is based on the advances made in optical communications technology and has benefited greatly from the developments made in this sector. This applies, for example, to optical fibers, light sources, couplings, connectors and receivers.

Despite the benefits of FOSs listed above, a large number of problems still have to be solved and it will be some time before a comprehensive range of applications emerges.

7.1 The structure of fibers

An optical fiber consists of an inner core of light-conducting material and a

80 Fiber optic sensors

surrounding cladding with a refractive index which must be less than that of the core. In practical applications, both materials are surrounded by a protective sheath. The light propagates through the core through total reflection at the core–sheath boundary. If the diameter of the light-conducting core is much greater than the wavelength of the light and if there is little change in refraction over an interval of one wavelength, then the propagation of the light waves can be described in terms of the so-called geometric optical approximation. However, there is the limitation that only a finite number of angles of inclination to the fiber axis are permitted. We then speak of the modes of the corresponding waveguide. The number of modes depends both on the wavelength of the light and on the geometry and distribution of refractive indices of the waveguide [134]. In a stepped-index fiber (Figure 7.1) the light can fall on one end of the fiber within a certain angle of acceptance $\pm\theta_A$. If this fiber end is in air, then

$$\sin \theta_A = (n_{CO}^2 - n_{CL}^2)^{1/2} = NA \tag{7.1}$$

where NA is the numerical aperture and n_{CO} and n_{CL} are the refractive indices of the core and the cladding, respectively. A typical fiber with a core diameter of 50 μm then possesses several hundred modes. In terms of the 'standardized structure constant' V[133], the number of modes is given by

$$M = V^2/2 \tag{7.2}$$

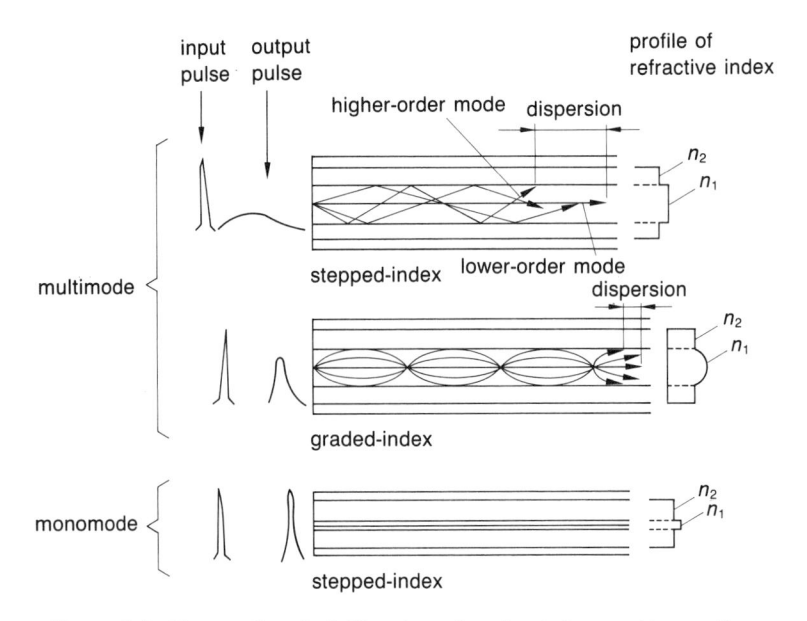

Figure 7.1 Types of optical fiber (*n*: refractive index; *r*: fiber radius).

where

$$V = r\pi(2/\lambda_0)\sqrt{n_{CO}^2 - n_{CL}^2} = r\pi(2/\lambda_0)NA \tag{7.3}$$

where λ_0 is the wavelength of light in a vacuum. In multimode optical fibers V has a value of between 30 and 50. From Figure 7.1 it is immediately clear that beams with different incident angles will require different times to travel a given distance. This is important in communications technology. A focused input pulse spreads during travel. This is known as *mode dispersion*, and it limits the bandwidth of the transmission signals in stepped-index fibers. In the example cited above, it is equal to 14 ns/km, which corresponds to a bandwidth of 70 Mbit km/s. The increased travel time of modes which are steeply inclined to the fiber axis can be partly compensated for if the refractive index declines constantly from its maximum value in the core to the cladding value. Such fibers are known as *graded-index fibers*. For a fiber with a 50 µm core and a numerical aperture of 0.22, mode dispersion is equal to only 0.035 ns/km.

Mode dispersion can be completely eliminated if the fiber is constructed in such a way that only one mode (strictly speaking, two orthogonal polarization states of a basic mode) can travel through it. This is possible at core diameters of less than 10 µm. Such fibers are known as *monomode fibers*. The structure constant mentioned above must satisfy the condition $V < 2.3$. From the technical point of view, it is much easier to produce multimode than monomode fibers [135]. While monomode fibers with core diameters of between 2 and 10 µm are manufactured exclusively from quartz (core and cladding), multimode fibers with core diameters of between 50 and several hundred micrometres can be produced from a variety of material combinations, for example:

- cladding and core of quartz glass;
- cladding and core of normal glass;
- plastic cladding and quartz glass core (so-called plastic-cladded fibers (PCFs);
- plastic core and cladding.

The correct type of fiber is selected for the intended application. Fiber optic bundles are also used for applications which require a particularly large light-conducting cross-section as well as a high degree of flexibility.

7.2 The classification of fiber optic sensors

The above-mentioned differences between multimode and monomode sensors mean that the two types have different areas of application. Although there is a vast body of literature concerning FOSs, their operation can be traced back to a small number of basic principles [136]. An initial distinction is that between *extrinsic* and *intrinsic*

82 Fiber optic sensors

sensors. In the case of extrinsic sensors, the fiber optics do no more than transport the light to the measuring point and the return signal to the receiver. The optical effect occurs in a different medium. Such sensors are usually multimode devices, although monomode sensors are occasionally employed. At the measuring point the incident light is modulated in accordance with the effect which is to be measured. In intrinsic sensors the measurand directly affects the transmission properties of the optical fiber. The optical effect occurs within the fiber and causes a measurable change in individual parameters such as geometric distortion or a change in refractive index or wavelength. Both multimode and monomode fibers are used. Both types of sensor are illustrated in Figure 7.2.

Another difference concerns the type of fiber. In multimode sensors the measured parameter influences the transmission properties of the optical fiber, that is to say, the intensity of the light transmitted by the optical fiber changes as a function of the measurand. External parameters acting on monomode sensors influence the phase of the lightwave which is passing through the fiber.

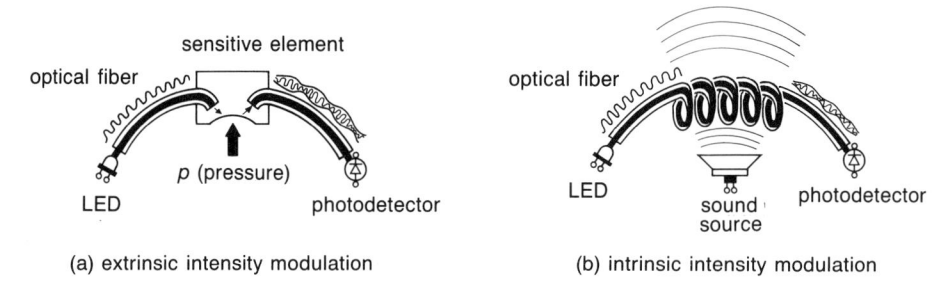

Figure 7.2 (a) An extrinsic and (b) an intrinsic sensor.

Table 7.1 Optical modulation and measurement methods

Modulation	Physical mechanism	Measuring method
Intensity	Change in light transmission through a change in absorption, emission or refractive index	Analog
Wavelength	Wavelength dependence of absorption, emission and refractive index	Comparison of the intensity of two wavelengths
Phase	Interference between signal and reference fibers or various propagation modes in a multimode optical fiber	Counting of stripes or phase measurement
Polarization	Change in the polarization tensor	Polarization analysis and comparison of amplitudes

Table 7.2 Optical effects for light modulation in fiber-optic sensors (n: change in refractive index, a: change in absorption, e: fluorescence emission)

Physical parameter	Optical modulation effect	
Mechanical force	Strain birefringent	n, a
Pressure	Piezo-optical effect	n
Deformation	Piezoabsorption	a
Change in density	Triboluminescence	e
Electric field	Electro-optical effect	n
Dielectrical polarization	Electrochromatic effect	a
Electric current	Electroluminescence	e
Magnetic field	Magneto-optical effect	n
Magnetic polarization	Faraday effect	n
	Magnetoabsorption	a
Temperature	Thermal effects	n, a, e
Change of chemical composition	Change of absorption	a
	Change of refractive index	n
	Fluorescence	e

Another important feature according to which sensors are differentiated is the type of modulation acting on the parameters of the light wave (Table 7.1). Intensity, phase, frequency, wavelength or polarization modulation are possible.

The terminological distinction between point sensors and distributed sensors is also important. In the first type, the reciprocal action of the physical effect and the light is localized. In the case of distributed sensors, the sensor is sensitive along its entire length or over certain discrete distributed sensor sectors.

For a summary of all the possible classifications, together with a large number of example applications refer to [137, 138]. Table 7.2 provides an overview of the effects which occur in the optical materials and fibers which are used in FOSs.

7.3 Applications

7.3.1 Multimode sensors

Currently, sensors based on multimode fibers still account for the majority of applications within the field of sensor technology. There are many potential fields of application, for example in process control [139–145], medicine [146–148], chemistry [149–150], the automobile industry [151], to name but a few. The simplest but most highly represented group of sensors within this category possesses only a single data transmission channel. These sensors consist of a transmitter–fiber–receiver combination. The microswitch (Figure 7.3) is an example of this. It provides a single binary datum. However, if the movable fiber can be continuously displaced then the same principle can be used to produce an analog-coded throughflow sensor. The hydrophone with moiré grid illustrated in

84 Fiber optic sensors

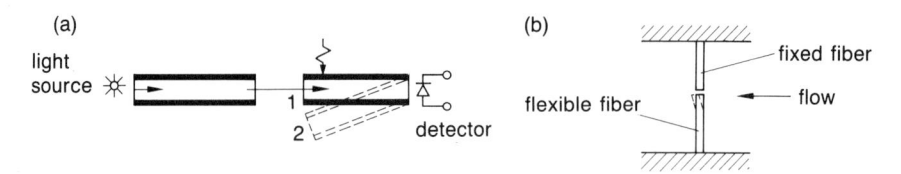

Figure 7.3 (a) Optical microswitch (position 1: closed; 2: open); and (b) its use as a flowmeter.

Figure 7.4 can be used to detect small pressures of the order of millipascals [153]. A movable grid is connected to the membrane on which the pressure acts. In the distance sensor, which exists in both dual- and single-fiber versions (Figure 7.5), the ratio of the output to input intensity varies as a function of the normed distance L/D (Figure 7.6). The b variant makes it possible to measure very small distances in the

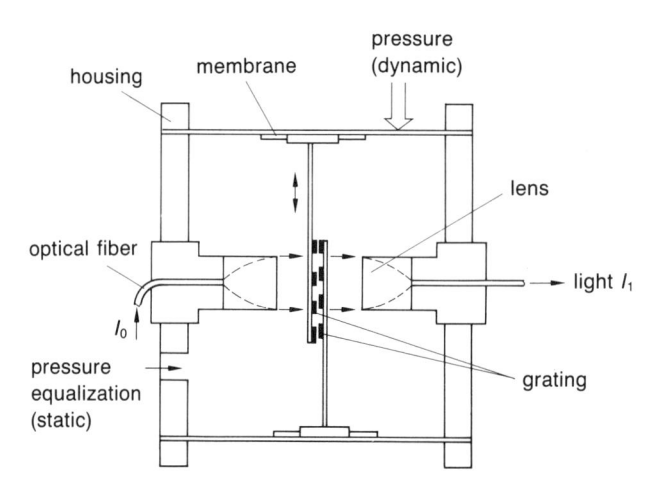

Figure 7.4 Hydrophone with movable grating [133].

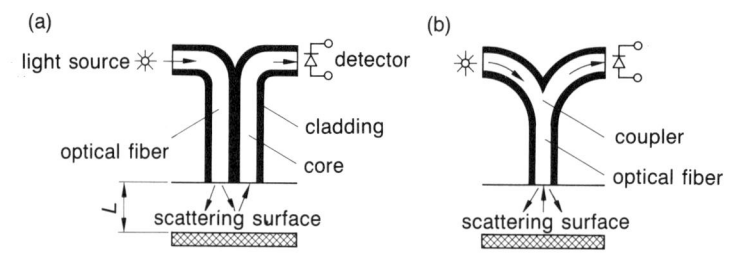

Figure 7.5 Distance measurement (a) using back-scattering or reflection in a double-fiber; (b) single-version fiber.

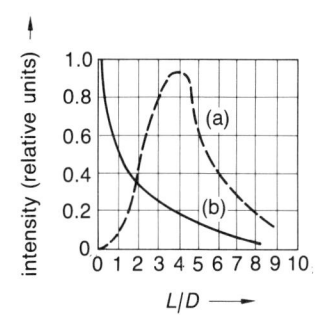

Figure 7.6 Received intensity as a function of the normed distance (*L*: distance; *D*: diameter of core; (a), (b) as in Figure 7.5).

range 0.1–1 mm. This configuration becomes a pressure sensor if the light-reflecting surface is a membrane on which the pressure acts. In principle it is also possible to measure turbidity, vibration, temperature and throughflow using this type of sensor [137]. The examples listed so far are all typical extrinsic sensors of which there is an enormous range.

Modification of the sensor illustrated in Figure 7.5 produces a filling level delimiter. This is formed by affixing a conical ground section or a prism to the end of the optical fiber (Figure 7.7). If the point of this is in the air then total reflection occurs. However, if it enters the liquid then the higher refractive index of the fluid causes some of the light to escape into it and the detector signal falls. In this way it is possible to measure fluid levels to an accuracy of 0.1 mm. The use of this type of delimiter is particularly useful in connection with highly flammable liquids, for example to indicate fuel tank overfilling or to detect leaks in pipelines. They can operate very reliably and, to a large extent, independently of distance. If, for example, a light–dark ratio of 10 dB is obtained at the detector and the threshold is optimally adjusted, losses in the fiber may vary by ±5 dB before a malfunction is reported. The range of variation this allows, together with the performance of 70 dB, permits considerable latitude in the design of this type of sensor. Of course the resolution is only 1 bit. Higher resolutions are theoretically possible if the

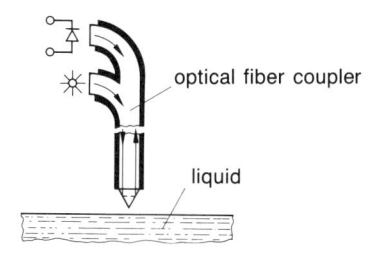

Figure 7.7 Design of a fiber optic filling level sensor.

modulation in the receiver and the evaluation of the received signal are not digital but analog. An example of this type of sensor is the so-called fluid refractometer which uses a U-shaped optical fiber (Figure 7.8) whose cladding has been removed in the curved section, thus exposing the light-conducting core directly to the environment. At a given optical fiber diameter and at a given wavelength, the number of modes capable of propagation is dependent on the difference between the refractive indices of the core and the cladding. As each mode transmits approximately equal quantities of light, the light output to be transmitted by the optical fiber is also dependent on this difference. If the optical fiber is in air then the values of n_{CO} and n_{CL} are determined by the fiber material (and therefore by the refractive indices of the core and the cladding). If the fiber is immersed in a liquid with the refractive index n_L, and if $n_{CO} > n_L > n_{CL}$, then the transmitted light output decreases. In extreme cases $n_F \geqslant n_{CO}$ and no light is transmitted. The simplest application for this sensor is as a filling level delimiter. An extended version can also be used as a refractometer, in which case the transmitted light output is brought into a direct dependence relation with the refractive index n_L. An absolute precision of 10^{-3} can be obtained and relative changes in refractive index of 10^{-5} have been detected. A linear characteristic exists for the range of refractive indices from 1.33 to 1.40. However, the operating temperature range is extremely small (20–45°C). In practical operation problems occur when liquid collects in the U-shaped area or if this becomes dirty. This category also includes the microbending pressure sensor (Figure 7.9). The principle underlying this sensor is that in regions where the optical fiber bends considerably (small bending radius), part of the light passing through the core is transferred to the cladding. This microbending effect, which is caused elasto-optically by mechanical strains, is used for sensitive pressure measurements in the 1 mPa range.

The fields of chemistry, biology and medicine offer an interesting range of applications for FOSs. Applications in the chemical field are extensively documented in the relevant chapter. To complete the preceding discussion I shall present the example of a pH sensor here (Figure 7.10). The chemical indicator is located at the end of the probe and its colour changes with the pH value of the surrounding liquid. This colour change is measured.

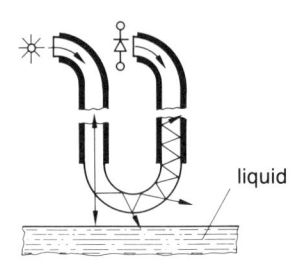

Figure 7.8 Liquid refractometer with U-shaped optical fiber [133].

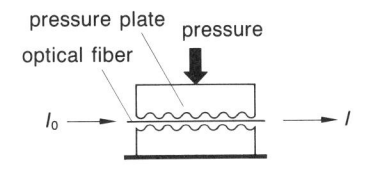

Figure 7.9 Pressure sensor based on the 'microbending' principle (I_0: input intensity; I: output intensity).

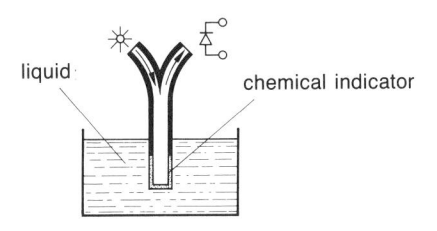

Figure 7.10 pH value sensor.

These last examples can be grouped among the analog-coding sensors with a single transmission channel. In principle, these sensors can provide a very high resolution. Given a signal-to-noise ratio of 30 dB at the receiver, which is easy to achieve in practice, it is possible to distinguish between 10^3 different x-values (this corresponds to a 10-bit resolution) provided that a high degree of constancy (± 0.004 dB) can be guaranteed for the fiber transmission losses. However, this is only possible in exceptional cases, for example when short, rigid, fixed fiber connections are used, as in the case of U-shaped refractometers, or when short times (less than 1 s) are involved, such as might be sufficient for the measurement of sound pressures.

Developments so far have shown that when analog encoding of intensities is to be performed, the variable losses which occur in the transmission path, in particular attenuation due to bending or at the connector, mean that a high degree of sensor sensitivity can be achieved only with considerable effort. However, the independence of the measurement from these losses, that is to say, path neutrality, is a decisive precondition for the long-term stability of the measurement of quasi-static measurands as well as for the operation of sensor networks. For this reason, much work is currently being conducted with the aim of developing path-neutral FOS designs. In most of these developments, the objective is the digital encoding of intensity or a signal coding of the frequency, time, phase or wavelength.

One way to achieve a considerable reduction of these unwanted influences is to transmit a reference signal along the fiber together with the measuring signal. This is a commonly used procedure in measurement technology. Although this makes the overall configuration more complicated, the effort is worthwhile since the reliability and accuracy of the sensor are considerably improved. The operating principle behind dual-channel, analog-coded sensors is illustrated in Figure 7.11. The primary luminous flux I_o is separated into two fluxes I_1 and I_2 by a filter. The measurand x modulates only I_1. I_1 and I_2 are passed along the same fiber to the evaluation unit where they are separated and the quotient of the detector fluxes is used as a measure for x. This is independent of the size of the transmission losses, provided that these are equal for I_1 and I_2. The moveable part A in Figure 7.11 could take the form of an absorber wedge and could thus be used for direction measurement [140].

There are a number of sensors, some of them in commercial use, which employ this principle. The best known of these is the ASEA temperature sensor [154], based on the principle illustrated in Figure 7.12. The actual sensing element is a piece of phosphor which is embedded in silicon adhesive at the end of the fiber. This phosphor is a temperature-dependent photoluminescent element, an Al/GaAs crystal. Light enters the fiber from an LED with a maximum intensity of approximately 750 nm (λ_0) and causes the phosphorus to luminesce. The luminescence is fed back along the same fiber and enters the detector element through a plug-in connector. Here, suitable optical separation techniques are employed to isolate two narrow spectral bands from the luminescent spectrum at 800 nm (λ_1) and 900 nm (λ_2). An electronic divider forms the quotient of the intensities, i.e. $I_1(\lambda_1)/I_2(\lambda_2)$. The monotonic temperature dependence of this luminescence–intensity ratio is used as the sensor effect. This type of sensor is designed to operate over a temperature range of 0–200°C at a resolution of approximately 0.1°C. The sensitive tip of such a sensor has a diameter of approximately 0.5 mm.

Other phosphors, such as materials doped with rare earth ions like $La_2O_2S:Eu^{3+}$, $Gd_2O_2S:Eu^{3+}$ or glasses doped with Nd^{3+}, are used as FOSs. Commercial sensors, such as the so-called Luxtron sensor, use this type of crystal [155].

Highly-developed FOS configurations also exist for high-temperature applications. Figure 7.13 illustrates a temperature sensor which makes use of a black body radiator at the fiber end. A sputtering technique is used to apply an iridium

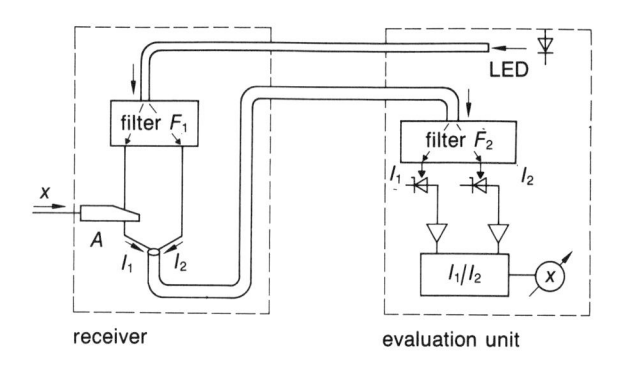

Figure 7.11 Fiber optic sensor with dual-channel analog coding [141].

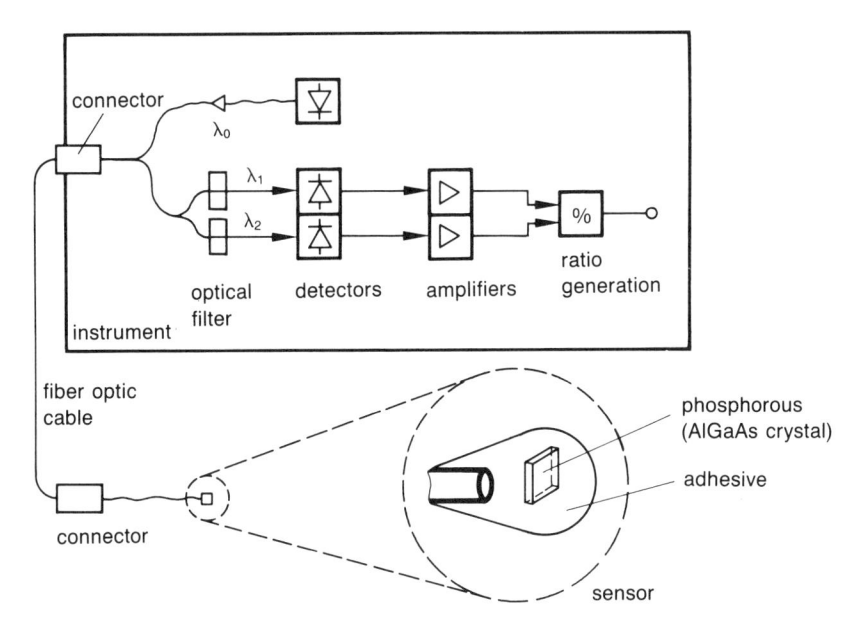

Figure 7.12 Schematic representation of a commercial temperature sensor with a temperature-dependent photoluminescent element at the end of the optical fiber [133].

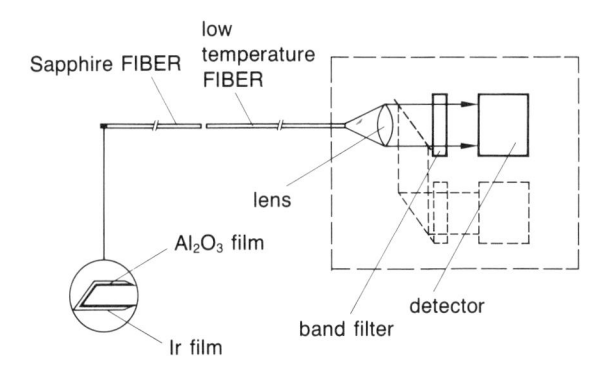

Figure 7.13 Accufiber high-temperature sensor with black body radiator [157].

film which acts as a black body to a temperature-resistant sapphire monocrystalline fiber. This film is protected externally by an Al_2O_3 film. The spectral density is measured as a function of the temperature within two narrow wavelength ranges ($\lambda_1 = 0.4$–0.5 μm; $\lambda_2 = 0.6$–0.7 μm). The sensor is suitable for making very precise measurements (measuring accuracy 0.05%) in the temperature range of 500–2000°C. Thanks to its small size and the low thermal conductivity of the sensor tip, this sensor possesses a wide dynamic range and is particularly well suited to taking measurements in gas flows. A sensor based on this principle has recently been used by the US National Bureau of Standards to define a temperature standard between 630°C and 1064°C. This testifies to its sensitivity, stability and high quality.

Apart from the two principles described above, there are other interesting designs which also modulate intensity or wavelength in order to measure temperatures. In this way it is possible to make use of the temperature-dependent behaviour of a variety of filters (birefringence, interference) for the modulation of light in a fiber optic system [158]. Thermochromous substances, whose colour changes with temperature, can be analyzed for the presence of certain wavelengths following the optical fiber transmission process [159]. Changes in the boundaries of absorption bands, for example in GaAs, are also used to measure temperatures over a small range.

One technique which avoids the need to compare intensities at two wavelengths is to measure the decay time of fluorescent emissions [156]. After pulse-type excitation, the intensity of the emitted radiation $I(t)$ decays with time in accordance with the relationship

$$I(t) = I_0 \exp(-t/\tau) \tag{7.4}$$

where τ is the decay time which is characteristic for the phosphor used (Figure 7.14). In the experimental realization described in [156], in which phosphate glass doped with Nd^{3+} was used, τ denotes the value of the half-life:

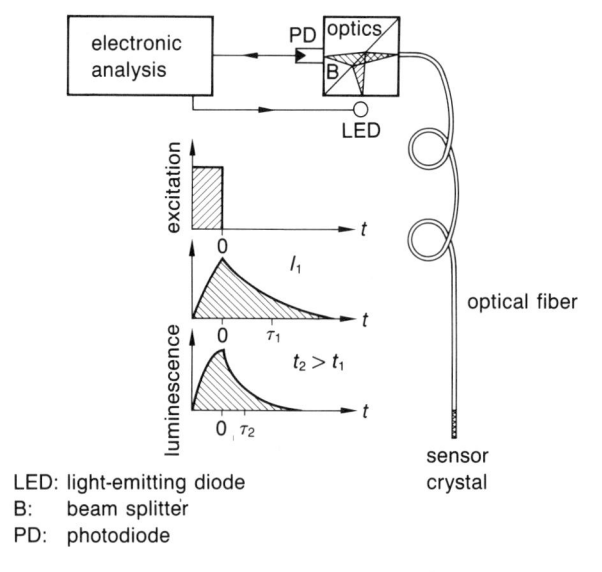

$$\tau = |t_2 - t_1|/\ln 2 \tag{7.5}$$

LED: light-emitting diode
B: beam splitter
PD: photodiode

Figure 7.14 Temperature measurement system based on luminescence decay time [162].

where $t_2 - t_1$ is the time difference between an arbitrarily determined value I_0' and $I_0'/2$. τ exhibits linear temperature dependence.

In a commercial application using a Cr-doped yttrium–aluminium–garnet (YAG) crystal the overall temperature-dependent decay time of the system is measured [162]. The weak luminescent light returned from the sensor is evaluated using phase-sensitive measuring techniques. After the decay time has been converted into a frequency-analog value it can be easily digitized. The sensor can operate at temperatures between $-50°C$ and $400°C$ at an accuracy of $0.5°C$. Response times shorter than 1 s can be achieved. In contrast to the quartz-based digital temperature sensors which will be discussed later, the sensor presented above can be used in environments where there is a risk of explosions and in environments subjected to high levels of electromagnetic radiation. This type of environment is frequently encountered in industrial applications.

An interesting variant of the sensor presented in Figure 7.12 produces a vibration or acceleration sensor [160]. In this version the end of the optical fiber is ground to form an obtuse angle on which is mounted a reflective spring body made of GaAs–AlGaAs produced using an anisotropic etching process (Figure 7.15). When accelerated, this bends and reflects an acceleration-dependent fraction I_1 of

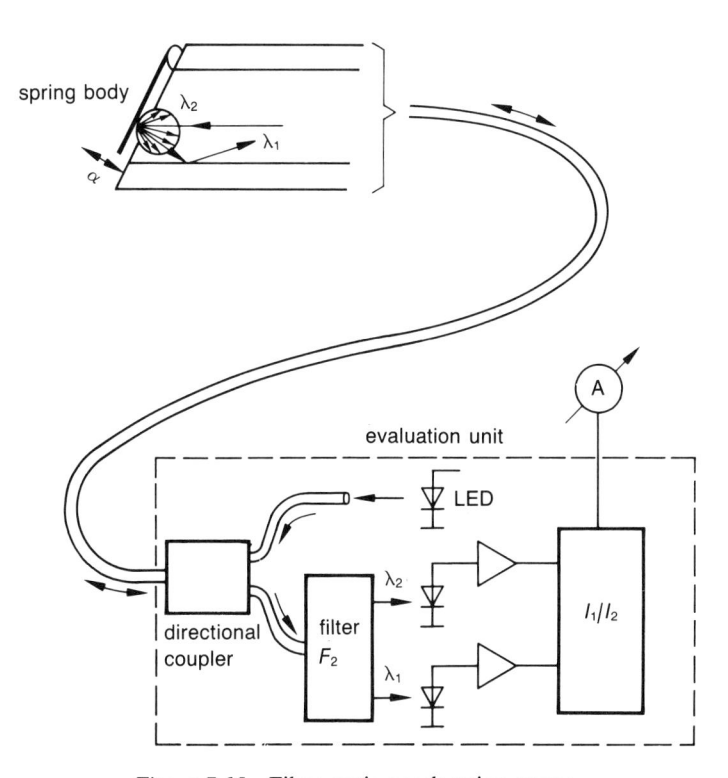

Figure 7.15 Fiber optic acceleration gauge.

the directed primary light source back into the fiber. At the same time it emits undirected fluorescent light, a proportion I_2 of which is returned to the fiber. The required information is derived from the quotient of the two intensities. This type of sensor, which is commercially available, provides a resolution of $0.05g$ with a dynamic range of 70 dB.

Sensors with a large number of transmission channels are particularly useful for industrial applications [141]. With such sensors it is possible, given M simultaneously usable channels each of which modulates as a binary digital optoswitch, to obtain a measuring resolution of $N = 2^M$ dots. The problem in construction consists of providing the necessary channel filters together with a modulator which can convert the measurand into a binary form. Figure 7.16 illustrates this principle at work in an angle sensor. The light falling on the sensor is spectrally divided into ten channels by a grating monochromator. This device is comparable to a wavelength multiplexer. The ten separated light fluxes pass through the modulator, a ten-track encoding disc. The light fluxes which pass through the modulator are combined and passed to the evaluation device where they are separated again for detection. The ten sensor signals form a ten-position binary number which gives the angle of the disc directly. The same principle is used in the

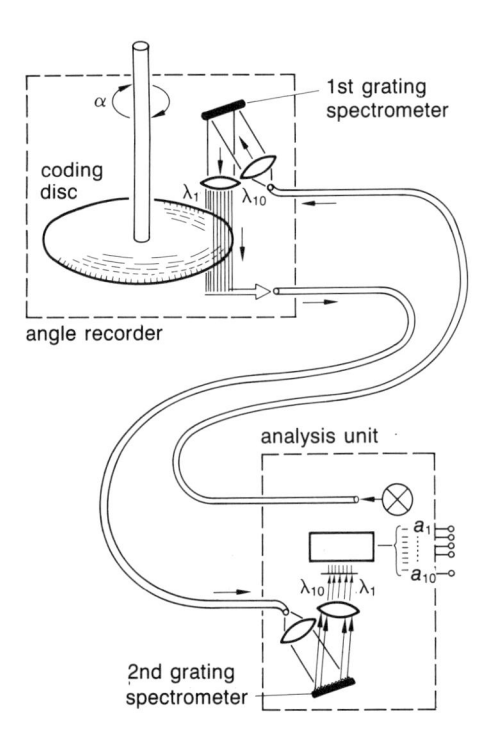

Figure 7.16 Fiber optic angle sensor with ten-channel digital encoding.

fiber optic position sensor described in [161]. Because of the digital modulation which they employ, both sensors possess excellent path-neutrality.

7.3.2 Monomode sensors

In this class of sensor the phase position or polarization of the light passing through the optical fiber is modified. Evaluation is performed using interferometric configurations. Although most of the sensors presented below are still at the development stage, interesting developments can be expected, especially in the field of high-sensitivity sensors. These include hydrophones, magnetometers, gyroscopes and distributed FOSs. It is clear that monomode sensors will never be available at prices comparable to those of multimode sensors and that they will not be employed for the same kinds of mass application. However, the high precision they can attain and their suitability for special applications gives them important advantages over multimode devices. Moreover, the development of new sensor fibers is making new types of sensor solution possible. Monomode sensors are already being used in medicine, aviation and space travel as well as in military technology. New applications can be expected in the fields of process measurement, chemistry and biotechnology.

7.3.2.1 Monomode sensors with phase modulation

In monomode optical fibers only one wave mode is capable of propagation. Thus the phase

$$\varphi(\beta) = \beta z = (2\pi n/\lambda_0) \cdot z \tag{7.6}$$

and the polarization along the fiber are unambiguously characterized. In this equation, β is the phase propagation constant, n is the effective refractive index of the wave mode and λ_0 is the wavelength of light in a vacuum. The mode of operation of monomode sensors is based on the dependence of the phase difference

$$\Delta\varphi(P) = \varphi_M - \varphi_R = (2\pi/\lambda_0)\{n_M(P) \cdot z_M(P) - n_R z_R\} \tag{7.7}$$

between a measuring optical fiber (subscript M) and a reference fiber (subscript R) for the measured parameter P. n_M and n_R are the refractive indices of the two optical fibers. They are usually identical and have lengths z_M and z_R, respectively. The measurand P only acts on the refractive index n_M and the length z_M. The optical length z_R and n_R of the reference path must be independent of P. The length of the measuring fiber z_M can be very great thanks to the low attenuation in the fiber. Consequently, extremely sensitive sensors can be constructed. However, since most physical quantities influence both n_M and z_M (Table 7.3) these sensors can be observed to possess a high degree of cross-sensitivity. This can only be avoided by compensating suitably via the reference fiber. This is a taxing task which is currently being investigated world-wide.

Table 7.3 Effect of measurands on n_M and z_M

Measurand	Effect on	
	z	n
Temperature	Thermal expansion	Thermo-optical effect
Magnetic field	Magnetostriction	Magneto-optical effect
Tensile stress	Elongation	Photoelastic effect
Electric field	Electrostriction	Electro-optical effect

The use of the phase modulation method requires not only a monomode fiber but also a coherent light source, for example a laser diode. Phase modulation has both positive and negative features. The main advantage is that a high degree of sensitivity can be achieved. There is no problem detecting phase changes of 10^{-4} rad using interferometric techniques. In principle it is possible to achieve resolutions of 10^{-9} rad [163]. The disadvantages are as follows [138]:

- the cross-sensitivity (for example, temperature and strain are not independent variables);
- the necessity of maintaining a high degree of stability in the coherent light source;
- the problem of connecting/coupling the monomode fiber to the coherent light source;
- guaranteeing reference stability;
- complex electronics.

In principle, there are many types of interferometer which are suitable for phase measurement tasks. Figure 7.17 is a schematic representation of the most important types of interferometer used in FOS technology. All the fiber optic interferometers which operate on this basis require, alongside a light source with a stable wavelength and a parameter-specific preparation of the optical fiber range, the use of monomodal directional couplers. These are primarily based on the so-called evanescent wave coupling between the cores of two adjacent fibers. From the technical point of view they are not easy to produce. Examples of these couplers,

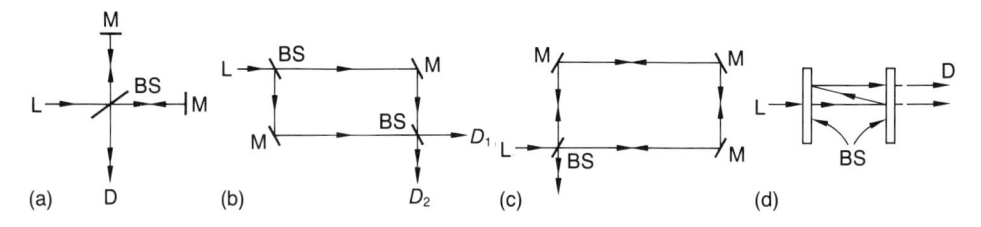

Figure 7.17 Dual-beam interferometer as proposed by (a) Michelson; (b) Mach and Zehnder; (c) Sagnac; (d) the multiple beam interferometer as proposed by Fabry and Pérot. L = light; D = detector; M = mirror; BS = beamsplitter [138].

which are produced by recrystallization, polishing, etching or, more recently, by integrated optics, are presented in [164–166]. I shall leave a more detailed description of these to the specialists. The same is true of phase modulators based, for example, on piezoceramics, or frequency shift devices, such as the Bragg cell, or the serrodyne principle [165]. They are also important components in the practical realization of fiber optic interferometer devices.

Before presenting the principles underlying the most important fiber optic interferometers and their application in greater detail, it is necessary to make a few basic preliminary comments about their mode of operation using the example of the Michelson interferometer (Figure 7.18). Light is emitted from a laser source and is divided into reference and signal beams in a beam splitter. The two beams can be represented as follows

$$\text{Reference beam} = A_R \exp[j\{\omega_L t + 2\beta x_R\}] \tag{7.8}$$

$$\text{Signal beam} = A_S \exp[j\{\omega_L t + 2\beta x_S\}] \tag{7.9}$$

where A_R and A_S are the amplitudes of the two beams and x_R and x_S are the paths covered by the beams between the reference and signal mirrors. β is given by

$$\beta = 2\pi n/\lambda_0 \tag{7.10}$$

where λ_0 is the wavelength in a vacuum, n is the refractive index of air. ω_L is the angular frequency of the light source. After passing through the interferometer arms the beams recombine coherently in the beam splitter. A current

$$I_D = \varepsilon\{1 + K \cos \varphi(t)\} \tag{7.11}$$

occurs at the photodetector. Here $\varphi(t)$ is the time-dependent phase difference between the interferometer arms:

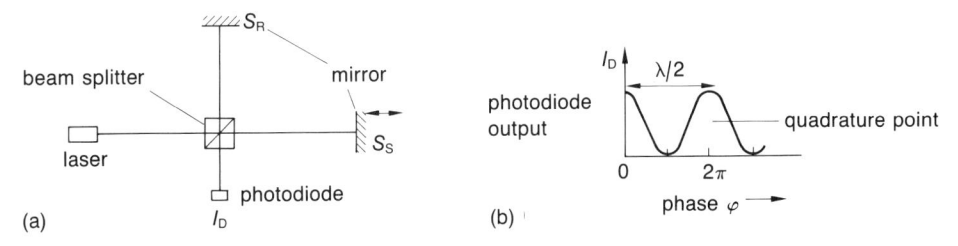

$$\varphi(t) = 2\beta \,|\, x_S(t) - x_R \,| \tag{7.12}$$

Figure 7.18 (a) Conventional Michelson interferometer; (b) transmission function presented as the dependence of the photon current I_D as a function of phase [164].

96 Fiber optic sensors

K is equal to 1 when $A_R = A_S$. This means that the two beams share the same intensity and the same polarity. ε is a constant referring to the input power. The value of I_D given by equation (7.11) is plotted in Figure 7.18(b). It can be seen that the output from the Michelson interferometer varies with the periodicity 2π which corresponds to a mirror displacement of $\lambda/2$. The maximum sensitivity to displacement $dI_D/d\varphi \sim \sin\varphi$ occurs at $\varphi = \pm n\pi \pm \pi/2$. This corresponds to a difference in the optical paths of $(2n+1)\lambda/4$. This is the so-called *quadrature position* or *quadrature signal*. This quadrature signal is used to achieve high levels of sensitivity and linearity of the transient response in sensor applications. This is achieved by applying the long-known methods of conventional optical interferometry, such as the homodyne or heterodyne methods [167] to fiber optic interferometers.

The example application shown in Figure 7.19 is a Mach–Zehnder interferometer [133]. Rapidly changing parameters such as sound or vibration act on the measuring element and are detected. In versions of this sensor that have been produced so far the lack of long-term stability has made it difficult to measure static parameters. In the first coupler the laser light is evenly divided between the measuring and reference fibers. The parameter P causes a phase change in the measuring fiber compared to the light travelling in the reference fiber. A second coupler causes the two lightpaths to interfere and the light intensities I_1 and I_2 are measured at the two detectors D_1 and D_2 which provide an identical, complementary signal. The differential signal, from which the amplitude noise of the laser is practically eliminated, is passed through a low-pass filter to an integrator. This sends a control signal to a piezoceramic transformer (PZT) around which the reference fiber is wound (phase shifter) and which is used to adjust the interferometer to its most sensitive operating setting, that is to say, a 90° phase shift is maintained between the measuring and reference paths (quadrature condition). Low-frequency phase changes, caused by drift in particular, are compensated for by the PZT

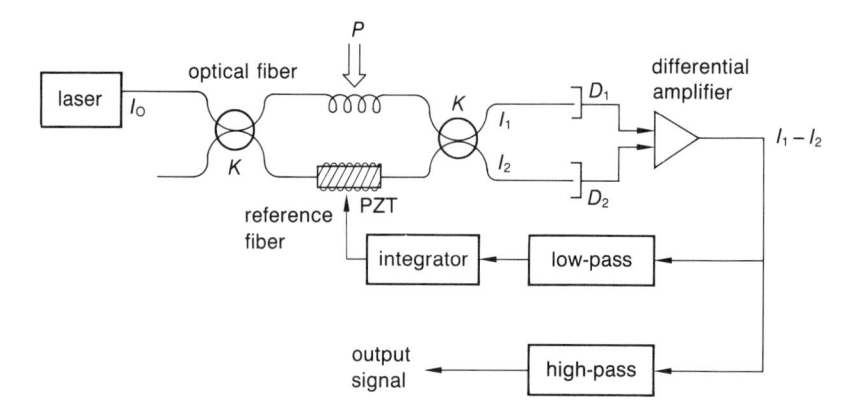

Figure 7.19 Fiber optic Mach–Zehnder interferometer with homodyne detection [133].

transformer which introduces an opposing phase shift in the reference fiber. Rapid changes in the intensity signal $I_1 - I_2$ caused by the measurand P are registered at the high-pass. The output of the high-pass thus provides the profile of the measured parameter P. This type of homodyne sensor, which operates by forming a differential signal, allows a phase measurement resolution of up to 10^{-7} rad/m for modulation frequencies above 1 kHz which are typical of the measurand. The first fiber-optic based hydrophone suitable for underwater operation employed the principle of the Mach–Zehnder interferometer and was able to measure sound pressures of 100 µPa at ranges of up to several hundred hertz. In acceleration or vibration sensors it is possible to achieve sensitivities of up to 10^{-2} µg (10^{-7} m s^{-2}) across a wide dynamic range [169]. The Mach–Zehnder principle is also of great interest for sensitive magnetic field sensors. In this case the measuring fiber is either surrounded or contacted by a magnetostrictive material. Figure 7.20 illustrates a number of possibilities. If a magnetic field is applied to the sensor the magnetostriction creates a longitudinal stress in the fiber which in turn causes a modulation of the light. It is possible to detect weak magnetic fields in the order of 10^{-6} A m^{-1} at frequencies of 100 kHz in a fiber 1 m in length. This type of magnetic field sensor is particularly well suited to measuring field gradients.

Mach–Zehnder sensors are of interest in military applications, such as submarine location, as well as in the field of medical diagnoses, non-destructive materials testing (NDT) and in geophysical exploration. Magnetic FOSs have a similar or superior performance to that obtained using the classic tools of fluxmeter, NMR or Hall effect. They are connected to the extremely high-sensitivity SQUID which, however, functions at low temperatures. In contrast, FOSs operate at room temperature and in EMI environments. Drift phenomena, sensor hysteresis and noise signals cause problems when low-frequency magnetic fields are present. However, complex electronics can be used to eliminate these unwanted effects and guarantee sensitivities of 10^{-5} A m^{-1} at 1 Hz [172]. It is also possible to detect alternating electric fields using FOSs based on the Mach–Zehnder principle [173]. To this end, the sensitive optical fiber is surrounded by an electret, Poly(VDF-TFE), whose piezoelectric properties are used indirectly to detect the field. Electrical fields stronger than 30 µV m^{-1} can be detected. Also of interest are developments in which a part of the passive cladding material is replaced by active material, for example by liquid crystal [174].

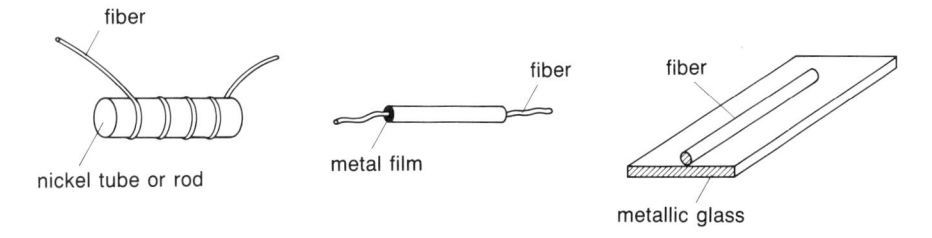

Figure 7.20 Basic configurations of magnetostrictive fiber optic sensors [171].

The fiber optic Michelson interferometer is used for high-resolution elongation measurements in the micrometre range and for temperature measurements. Where high-resolution temperature measurement is required, Michelson configurations provide a resolution of 10^{-3} K with a measuring range of 0–250°C [170]. Mach–Zehnder and Michelson interferometers are dual-beam interferometers. The principle behind the classical Fabry–Pérot interferometer is that of multiple beam interference, and this principle is also found in FOSs. The actual sensing element is a fiber optic Fabry–Pérot resonator (FPR). As illustrated in Figure 7.21, this may consist of a piece of monomode fiber of length L on the flat ends of which highly reflective dielectric films are formed. In this optical resonator the transmission rate for a given wavelength λ_0 may be large or small depending on the resonance condition, which itself is dependent on the length of the optical path nL. Maximum transmission occurs when the phase difference $\bar{\varphi}$ between the incident and reflected wave at surface 1 is given by

$$\bar{\varphi} = (2\pi L/\lambda_0)2\pi \tag{7.13}$$

Any change in the length of the optical path results in a phase change. The sensor effect is created by the fact that the parameter P which is to be measured (e.g., temperature, magnetic field, elongation) modifies the length of the optical resonator:

$$d(nL) = L\ dP\{(n/L)\partial L/\partial P + \partial n/\partial P\} \tag{7.14}$$

As an example, Figure 7.22 shows a train of transmission peaks for a 30 mm long Fabry–Pérot sensor as a result of changes in temperature. The peaks are repeated at intervals of approximately 0.5°C, which corresponds to $d\varphi = 2\pi$. The effect of the measurand is either recorded by counting the succession of transmission peaks or compensated for by regulating the wavelength. One problem associated with this type of sensor is that they have to be recalibrated with each start-up as otherwise only relative values can be measured [133].

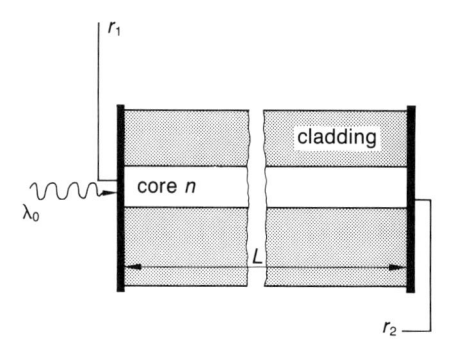

Figure 7.21 Fiber optic Fabry–Pérot resonator.

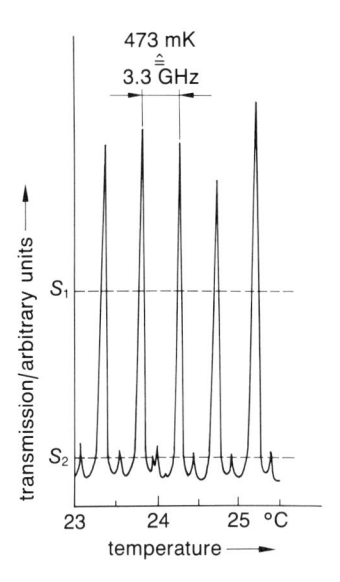

Figure 7.22 Transmission in a fiber optic Fabry–Pérot resonator as a function of its temperature [133].

One of the most promising areas of application of a fiber optic monomode sensor is in optical gyroscope systems. These can be used in the inertial navigation systems of aeroplanes or ships. The great advantage of this type of fiber optic based rotation sensor over mechanical types lies in the fact that it requires no rotating mass or moving mechanical parts. Although the first commercial fiber optic gyroscopes have now appeared, development is still continuing at full speed [175–177]. The optical gyroscope is based on the Sagnac effect which was discovered in 1913 and which is illustrated in Figure 7.23. A beam of light can travel around a closed circle

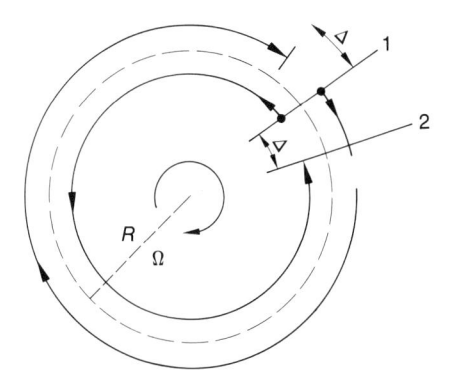

Figure 7.23 The Sagnac effect.

either clockwise or anticlockwise. For the sake of simplicity, let us assume that the medium is a vacuum so that the light travels with the speed of light in a vacuum c_0. The system is at rest. If a beam of light is now sent simultaneously in both the clockwise and the anticlockwise direction from point 1 then both will arrive back at point 1 at the same time. If the system is now moved clockwise at the angular velocity Ω, as illustrated in Figure 7.23, then point 1 is moved to point 2 while the light is circulating. The light travelling in the anticlockwise direction arrives at point 2 at a time when the beam circulating in the clockwise direction is still 2Δ away from point 2. As a result there is a phase difference between the two beams after completing the ring which a complicated calculation shows to be

$$\Delta\Phi = (8\pi A/\lambda_0 c_0)\Omega \tag{7.15}$$

where A is the area enclosed by the fiber optic ring. Equation (7.15) shows that the angular velocity Ω is linearly proportional to the phase shift $\Delta\Phi$. It is important to enclose as large an area as possible so that rotational speeds of 1/100 or 1/1000 of the earth's speed of rotation can be measured precisely.

The availability of monomode fibers with low attenuation has made it possible to realize a large enclosed area in a small volume. This is done by, for example, winding a fiber of length 1000 m into a coil with a diameter of 10 cm. The enclosed area of 3180 windings is then traversed [175].

The principle of the fiber gyroscope is easily understood. However, putting it into practical use raises a whole series of problems of which only a few will be mentioned here. The Sagnac effect is a non-reciprocal effect, that is to say, it depends on the direction of the beam. This is illustrated in Figure 7.24. The wave travelling in the clockwise direction is transmitted to the beam splitter twice while the anticlockwise beam is reflected twice. This causes an unwanted phase offset. This problem is solved by incorporating a second beam splitter (Figure 7.25). This configuration is known as a *reciprocal construction*. The polarizer is used to maintain a defined polarity state. In principle, two modes exhibiting orthogonal polarity states of the basic mode can propagate in a monomode fiber. In order to eliminate measurement errors caused by birefringence or mode conversion a defined

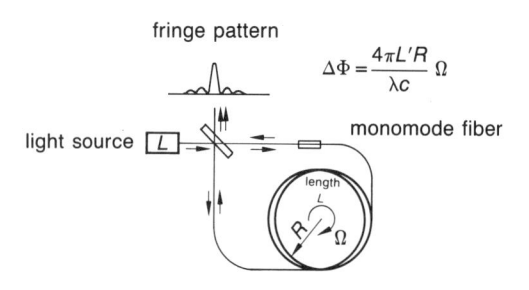

$$\Delta\Phi = \frac{4\pi L'R}{\lambda c}\Omega$$

Figure 7.24 Principle of a fiber-optic gyroscope [175].

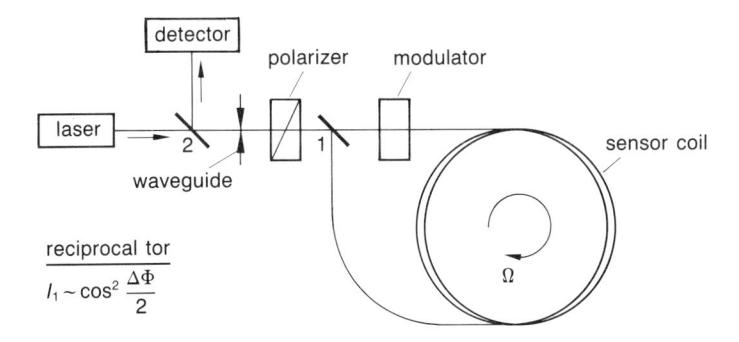

Figure 7.25 Minimum configuration of a fiber optic gyroscope [175].

polarity state must be guaranteed in the interferometer. The task of the modulator is to introduce a stable phase offset of $\pi/2$ between the two beam portions. This is necessary at low speeds of rotation because although the cosine-shaped intensity curve is maximum at zero speed of rotation, the sensitivity to phase changes is at its minimum at this value. Phase and frequency modulation procedures are used [175].

Laboratory models using discrete optical components have yielded excellent measuring results. Problems still to be solved relate to the long-term stability of the devices and increasing the resolution. In turn, these questions are closely bound up with the technological realization of the fiber gyroscope. Today it appears that it is the following technologies which will make the fiber gyroscope a competitor of the mechanical gyroscopes:

- the use of polarization-maintaining monomode fibers [178];
- the miniaturization of interferometer construction through integrated optics or with fiber optic components (all-fiber solution); or
- hybrid construction using integrated optical, fiber optical and micro-optical components.

The aim of the all-fiber solution is to produce the entire optical construction of the fiber gyroscope from a continuous piece of monomode fiber. An alternative to this is the use of integrated optical circuits based on lithium niobate. The basic structures are monomode strip fibers made of $LiNbO_3$ which can be produced by diffusing Ti atoms in $LiNbO_3$ substrates. The electro-optical and acousto-optical properties of the various crystal cuts of $LiNbO_3$ crystals are also a precondition for the ability to use this material to produce not only passive structures but also active ones such as optical switches, phase and frequency modulators. Figure 7.26 gives some examples of integrated optical chips for the fiber gyroscope. Integration has advanced to the extent where the latest commercial fiber gyroscopes with phase modulators consist of only four components, namely the laser and detector modules, the sensor coil and the integrated optical chip [176].

102 Fiber optic sensors

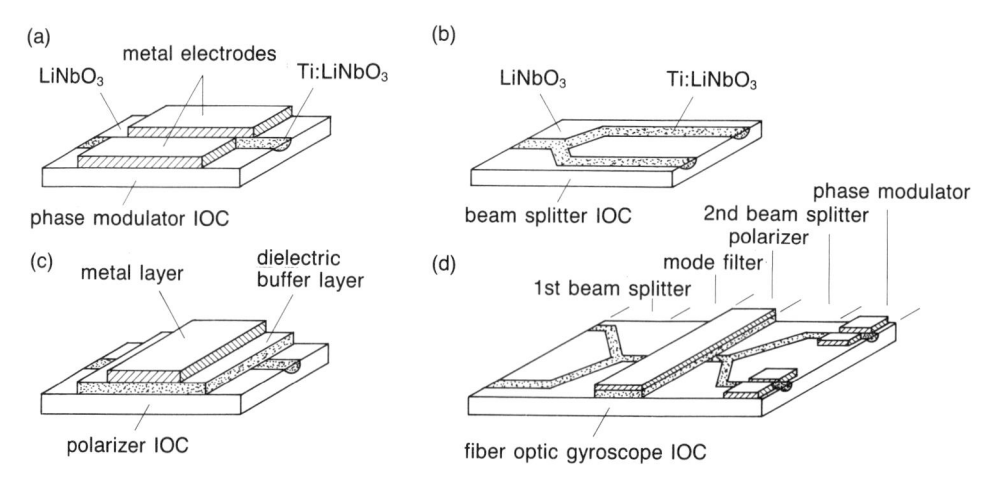

Figure 7.26 Integrated optical chips for the fiber optic gyroscope [166].

Using today's fiber optic gyroscopes it is possible to obtain resolutions of 10^{-2} degrees/h and thus to approach the values achieved by the most sensitive mechanical gyroscopes. Fiber optic gyroscopes are particularly well suited for use under conditions of high dynamic strain. If the drift behaviour is improved further, then these fiber optic devices will become genuine competitors to the mechanical gyroscopes.

7.3.2.2 Monomode sensors with polarization modulation

Birefringence has long been used for the analysis of static or dynamic stresses in optical structures and polarimeters have been used to determine the rotation of the plane of polarization of optically active substances. Both are now employed in monomode fibers which have an unavoidable weakness or which special processing has caused to exhibit a high level of birefringence. This means that the refractive indices n_x and n_y and thus the constants of phase propagation $\beta_y = 2\pi n_y/\lambda_0$ and $\beta_x = 2\pi n_x/\lambda_0$ for the orthogonally aligned directions x and y are different. For two linearly polarized waves in the x or y direction there is a phase difference after traversing the fiber length L of

$$d\Phi(L) = (\beta_x - \beta_y)L \tag{7.16}$$

which is known as the *differential phase* between the orthogonal modes. This value increases with the difference between the refractive indices $n_x - n_y$, that is to say, with the increasing birefringence of the optical fiber. A measured parameter P

acting on the optical fiber can affect not only the refractive indices n_x, n_y but also the length L. Thus for the differential phase

$$d\Phi(P) = 2\pi/\lambda_0 \{n_x(P) - n_y(P)\} L(P) \tag{7.17}$$

Sensors which detect the change in birefringence in normal monomode fibers as a function of temperature exist as laboratory prototypes [180]. However, their sensitivity is considerably lower than in the interferometric, phase-modulated method. This situation can be improved with more complex optics and electronics. More important for the development of new polarimetric sensors, however, are advances in the materials sector. These developments have made it possible to produce fibers with special polarization characteristics. These are, for example, fibers with negligible birefringence or fibers with a high linear birefringence. The latter category can be subdivided into 'polarization-maintaining' fibers and 'polarizing' fibers (or 'single polarization' fibers).

The use of optical fibers with a high birefringence has made it possible to develop new polarimetric sensors with sensitivities which approach those found in phase-modulated configurations [182]. The sensor and the processing electronics can be located far apart. This is not possible using conventional monomode fibers since additional birefringence effects can occur on the connection path and these can lead to undesirable side-effects. An example of this type of sensor with a polarization-maintaining fiber is given in Figure 7.27. The actual sensing element is a short piece of fiber with birefringence which is fused with a similar fiber of great length. The relative orientation of the intrinsic axes of the two fibers differs by 45°. The long fiber serves as the input and output connection for the system. Light from a linearly polarized source stimulates only one intrinsic mode in the input fiber. At the junction both intrinsic modes are stimulated in the sensor fiber. After they have been reflected at the silvered end they recombine at the junction. Elliptically polarized light appears at the output. This light is determined by the birefringence in the sensor fiber. Any change in this birefringence is recorded by determining the intrinsic mode which was not present in the input fiber signal. This system is suitable for the measurement of both temperature and displacement. If suitable signal processing electronics is used a very high degree of resolution can be obtained (e.g., a resolution of 10^{-5} rad in the case of phase measurement).

Fibers with low linear polarization or, better still, polarization-maintaining fibers are of considerable interest for the detection of magnetic fields and electric currents. If a magnetic field acts on this type of fiber, this causes a rotation of the plane of polarization in the optical fiber. This is known as the Faraday effect:

$$\alpha = V \int_L H \, dx \tag{7.18}$$

where V is the so-called Verdet constant, H is the magnetic field and L is the length

104 Fiber optic sensors

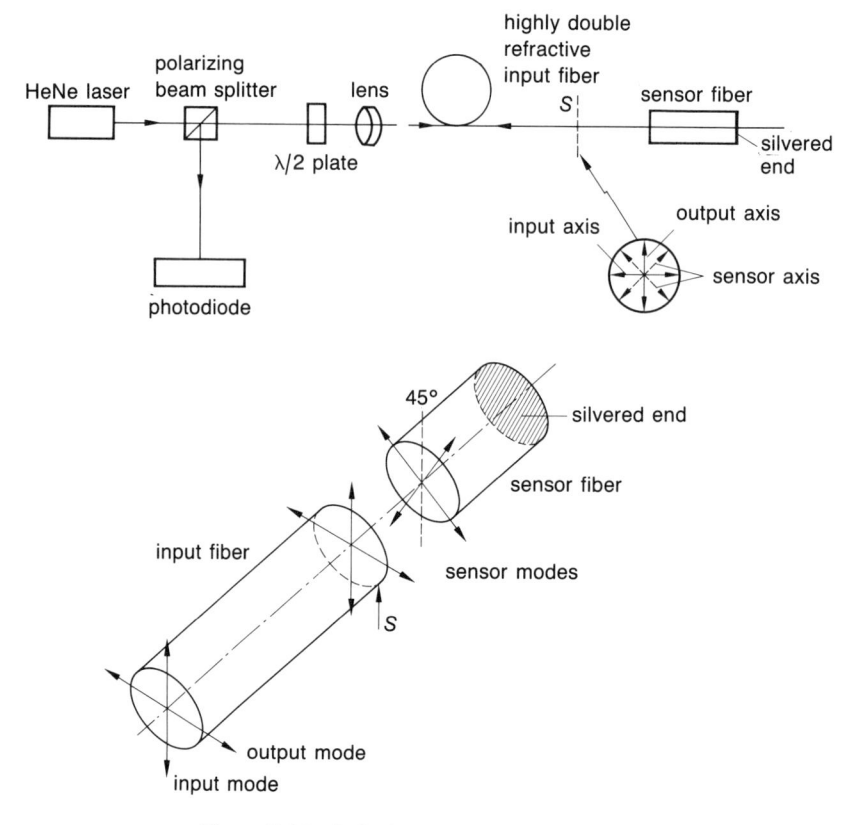

Figure 7.27 Polarimetric 'telescope' sensor.

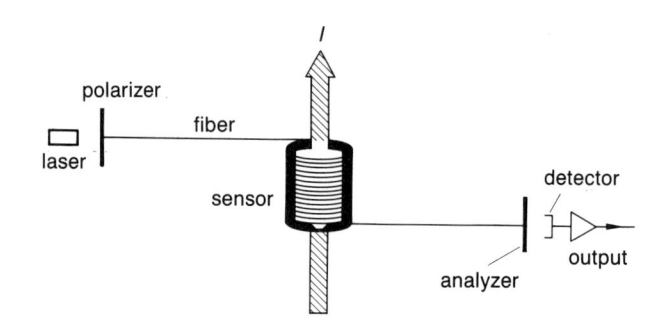

Figure 7.28 A magnetic field sensor which uses the Faraday effect.

of the fiber. If the monomode fiber is looped around the electrical conductor (Figure 7.28), then

$$\alpha = VNI \tag{7.19}$$

where N is the number of windings of the optical fiber coil. This type of Faraday sensor has been successfully employed for current measurement in high-voltage lines. Currents of 0.2–2000 A at frequencies of between 50 Hz and 100 kHz can be detected [171]. Intensive research is being directed at materials with a high Verdet constant with the aim of developing more sensitive Faraday sensors [184]. Fibers with a high level of angular birefringence are also being developed. This would considerably increase the sensitivity of current or magnetic field sensors as the effect of interference would no longer be noticeable.

7.3.3 Distributed fiber optic sensors

A specific area of application for optical fibers in sensor technology is the sensing of the effect of measurands over continuous or even discrete distributed fiber ranges. Whereas in the case of gyroscopes or optical magnetic field sensors the integrity of the information is maintained over considerable fiber lengths, distributed sensors make it possible to determine differential information. A locally triggered signal is received and the spatial distribution of the measurand values has to be determined. There is a wide range of possible applications such as the observation of stress distribution in buildings, pressure vessels, aircraft and oil pipelines, and of the temperature distribution in transformers, generators, heaters and reactors. Such sensors can also be used in alarm systems or in security zones.

This type of sensor is subdivided into two types. When measurements are performed at discrete, predetermined points of the fiber it is customary to speak of *semi-distributed* systems which are frequently also known as *multiplexed* systems. When measurements are taken continuously along the whole length of the fiber, the term *distributed sensor* is used [185].

An example configuration for a multiplexed system is given in Figure 7.29. The pulse from a laser source reaches the main fiber which simultaneously acts as a delay element. Part of the energy of the light pulse is split off at the coupling points and is fed to the sensor. This can work in either transmission or reflection mode. In reflection sensors the reflected light arrives at the receiver. A series of pulses appears, each of which corresponds to an individual sensor. The amplitude of each pulse corresponds to the state of its associated sensor. The duration of the transmission pulse must be smaller than the delay time between two sensors. However, it is not easy to put this principle into practice. When a large number of sensors are employed, problems relating to the required bandwidth and the power supply occur [186]. Other methods, such as wavelength multiplexing or techniques for modulating the source, partially avoid this problem. Although much development work still has to be done, the first applications for this type of sensor system are

106 Fiber optic sensors

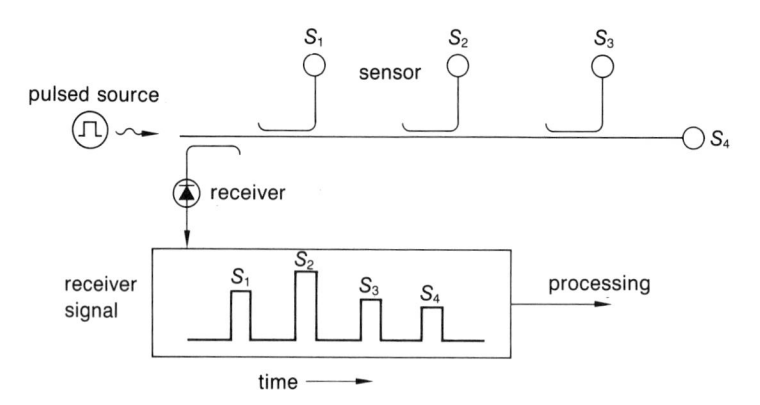

Figure 7.29 Time-multiplexed sensor system [165].

beginning to emerge. Sensor arrays containing up to 100 sensors are being used for hydrophones. The so-called vector scanner processes amplitude and directional information. The measurands for this type of device can be magnetic fields, acceleration and angular velocities [186]. Although technically difficult to realize, a type of sensor which is of great interest is one along which Fabry–Pérots are extended [187] or in which bulk gratings are inscribed (Melz in [188]). One of the first temperature systems was a semi-distributed configuration which was known as a differential absorption distributed thermometer [189]. A number of glass ruby plates (approximately 0.25 mm) located at the measuring points are connected using monomode fibers. The temperature-dependent absorption limit of the ruby glass is used for the measurement, and the Rayleigh light which is scattered back is analyzed at regular intervals. One of the main disadvantages of this configuration is that the absorption of each plate attenuates the signal and that the number of measuring points is consequently limited to ten.

Distributed sensors have opened up interesting avenues of development. They make use of the internal sensing properties of optical fibers in order to detect effects along the whole length of the fiber. Currently the greatest attention in the field of measuring technology is being paid to the temperature measurand. There are two reasons for this: first, the great need in many different fields to be able to measure spatial temperature distributions; and second, the relatively slow temperature response time in many applications which results from the high thermal capacity usually found in the systems under observation. This reduces detection problems.

Alongside temperature sensors, pressure and strain sensors also have an important role to play. Less important areas of application for distributed sensors are the measurement of magnetic and electric fields.

Most distributed sensors are based on reflectometric methods, that is to say, discrete reflections along the fiber or scattering effects in the medium itself are exploited. Optical time domain reflectometry (OTDR) is a technique in which a short, high-energy optical pulse enters the fiber and the energy scattered back along the fiber is observed as a function of time. This time is then proportional to the path through the fiber along which the scattered light is returned (Figure 7.30). In optical fibers most of the energy which is scattered back is caused by Rayleigh scattering resulting from low refractive indices. These are caused by fluctuations in density which are thermally induced in the fiber at high temperature and which are frozen into the fiber structure as the liquid glass is cooled. The elastically scattered light which these density fluctuations cause thus changes by only a relatively small amount with the temperature of the glass.

The situation is quite different with liquids, in which scattering results from fluctuations in the refractive index and exhibits a significant temperature coefficient. This fact is used to measure temperature distributions along fibers with fluid cores [190]. Figure 7.31 illustrates the principle underlying the overall configuration and the OTDR signal. A linear scattering coefficient of 0.4% $°C^{-1}$ is obtained up to 80°C. A temperature resolution of ±0.2°C can be obtained over several hundred metres at a path resolution of 2 m. The maximum operating temperature is approximately 160°C, as above this temperature the refractive indices of the core and the fiber are equal and the light-transmitting properties of the fiber are lost. Moreover, only solid-core fibers with the above-mentioned low temperature sensitivity can be used. Despite the interesting properties of liquid-core fibers, they have not yet achieved widespread use because they are difficult to use in many applications and because of disadvantages such as the need to supply a compensating tank, the influence of impurities, and high dispersion levels in the stepped-index version [191].

Many exciting new possibilities have been opened up by fibers which are doped with rare earths such as Nd^{3+}, Er^{3+} and Ho^{3+}. They possess a temperature-dependent absorption spectrum together with the properties of OTDR [192]. Temperature resolutions of 1°C at intervals of 1 m can be obtained with cable lengths of over 100 m [188]. Improved fiber production techniques can be expected to result in a huge expansion of this sector.

A technique which is still undergoing development but which may be of considerable significance in the future is known as *distributed anti-Stokes Raman*

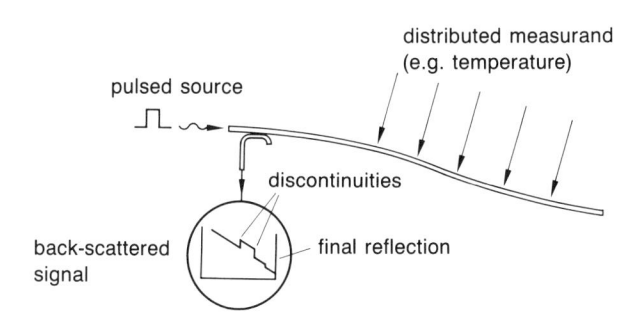

Figure 7.30 Optical time domain reflectometry.

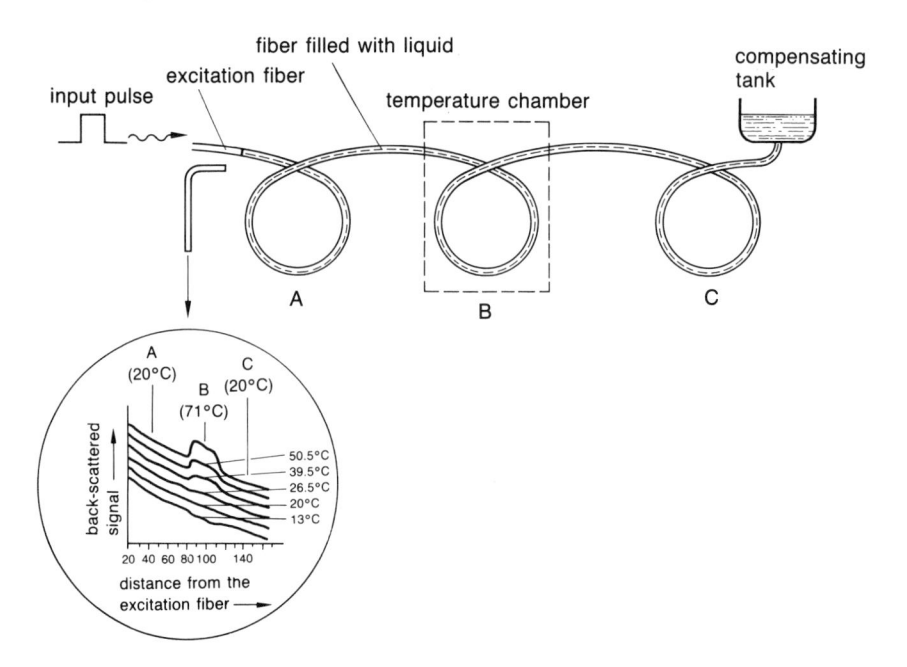

Figure 7.31 Scatter-dependent temperature measurement [185].

thermometry [193]. This differs in a number of points from the techniques described above. It employs an optically non-linear effect, the Raman effect, and operates as a distributed temperature sensor independently of the fiber material. This means that it can be used in fibers which have already been installed, for example in communications systems. Another advantage is that it uses absolute temperatures, so that no calibration is necessary. The method employs the spontaneous Raman effect. In this, molecular movements such as rotation or vibration cause the appearance of two weak lines in the spectrum, namely the Stokes and anti-Stokes lines (Figure 7.32). These lines can be explained in terms of quantum physics. For sensor technology it is important that for the intensity ratio

$$R = I_s/I_a \tag{7.20}$$

of the two lines at the wavelengths λ_s (Stokes) and λ_a (anti-Stokes) the following applies

$$R(T) = (\lambda_s/\lambda_a)^4 \exp(-h\nu/kT) \tag{7.21}$$

where ν is the frequency of the incident light, h is Planck's constant and k is Boltzmann's constant. The absolute temperature T is the only unknown and can be determined. This distributed sensor apparently has only one disadvantage, namely

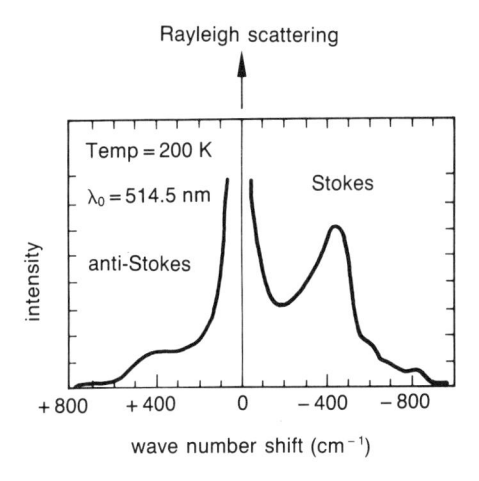

Figure 7.32 Raman spectrum of a Ge-doped silicon fiber [191].

that the anti-Stokes signal which is scattered back is 20–30 dB weaker than the Rayleigh signal. This means that a longer integration time is required if an adequate signal-to-noise ratio is to be achieved. This period amounts to 100 s at a laser repeat frequency of 40 kHz [185].

The first results of temperature profile measurements at fiber lengths of over 100 m are available. Resolutions of 1 K at intervals along the fiber of 1 m have been obtained. However, the demands on the receiving equipment are considerable and it is only worth using this type of sensor where other sensors cannot be employed. Apart from temperature, strain and power or pressure are parameters which it is very sensible and necessary to measure using distributed systems. The elasto-optical effect induced by the action of these measurands on the fiber causes either a change in the refractive index or, in the case of fibers with special polarization properties, changes in polarization. Figure 7.33 presents the concept of polarization OTDR (POTDR) [194]. A force F acts on a short length of fiber of length Δl. This causes a change in the polarization properties of the scattered light which is detected in the time-resolved received signal. The same principle can be used to detect electric fields.

The field of distributed FOSs is still only at an embryonic stage. With the development of integrated optics, of new fiber materials for both core and cladding, and with progress in the examination of non-linear optical effects such as the Zeemann or Stark effect, it will be possible to overcome economic limitations and open up new fields of application.

7.3.4 Other selected fiber optic sensor techniques

There are a large number of reciprocal surface mechanisms. The importance of two of these mechanisms, which have in fact been known for many years, has increased

110 Fiber optic sensors

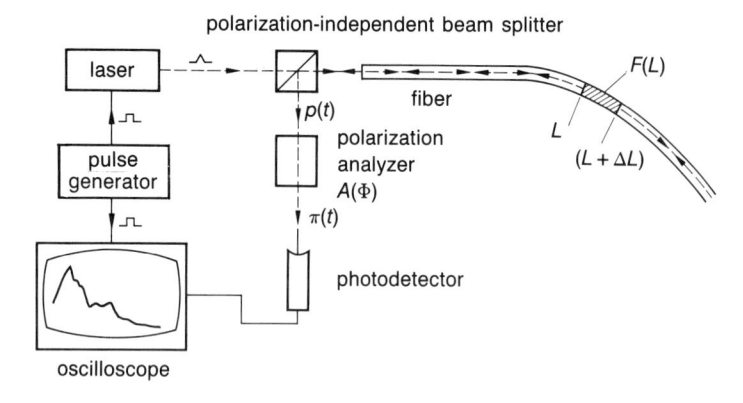

Figure 7.33 Polarization OTDR [185].

in recent years: plasmon absorption and variants of the evanescent wave which occur at conditions of total reflection in thin optical media [195]. I shall illustrate the latter using the example of a frequency-analog quartz pressure sensor [196]. Figure 7.34(a) presents a diagrammatic view of the complete configuration. A quartz crystal is mounted on a rod which is supported at one end and which is made to resonate by an electric source. If the end of the rod moves from its zero position the resonant frequency of the quartz changes because of the altered configuration of mechanical strains. The optical fiber is used to detect these changes. This is illustrated in Figure 7.34(b). The polished end of the input fiber is cut at an angle θ to the axis of the optical fiber. θ is greater than the critical angle θ_c. If the optical fiber and the quartz are sufficiently close to one another (<1 µm) then part of the light energy from the optical fiber can be transferred across the air gap from the optical fiber to the quartz. This is known as *frustrated total internal reflection*. This

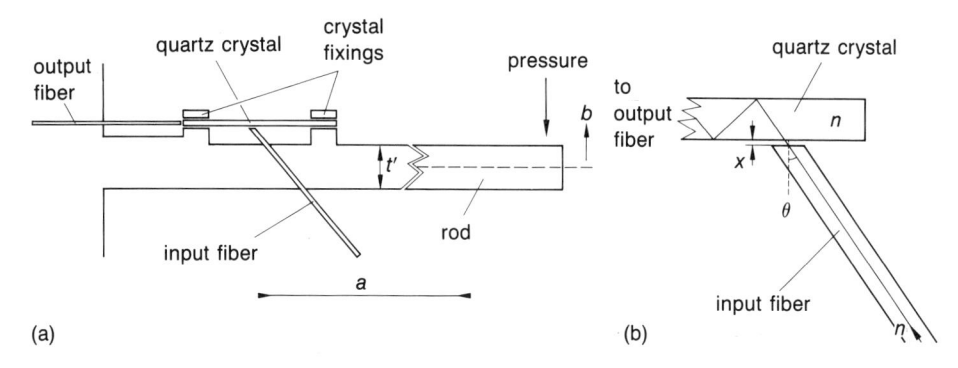

Figure 7.34 (a) Bending bar with attached resonating quartz and optical fiber used as a pressure sensor [196]; (b) representation of the coupling of optical fiber and quartz by an evanescent wave.

Applications 111

phenomenon is analogous to the tunnel effect in quantum mechanics. Transmission across the air gap is determined by the following relation [197]:

$$T = 1 - \{(z^2 + \delta^2)\ [(z^2 - \delta^2)^2 + 4z^2\delta^2\ \coth^2(\beta/2)]^{-1}\} \tag{7.22}$$

where

$$\beta = 4\pi x/\lambda (n^2 \sin^2\theta - 1)^{1/2} \tag{7.23}$$

$$z = 1/n \cos\theta \tag{7.24}$$

$$\delta = -(n^2 \sin^2\theta - 1)^{-1/2} \tag{7.25}$$

for light which is polarized perpendicular to the incident plane. n is the refractive index of the optical fiber and the quartz, x is the width of the air gap, λ is the wavelength in the dense medium. Figure 7.35 illustrates transmission across the air gap as a function of the gap width. The transferred light is linked into an output fiber at the quartz. As the amplitude of this light changes with variations in the distance between the input fiber and the quartz, an intensity-modulated signal with the frequency of the quartz resonator is output. In this way small changes in pressure can be detected.

The existence of this evanescent wave makes further sensor applications possible. For example, the cladding of the fiber core can be selectively modified by etching, implantation or other treatment in a way that effects the evanescent wave in the cladding. This can modify both the intensity and polarization of core transmission. An innovative idea is presented in [198]. Instead of the fixed cladding of the fiber core, the authors propose a gas mixture, in this particular instance consisting of methane and air. The absorption or decay time depends on the methane concentration. The result is a gas sensor. Other resonance methods using fiber optic sensing have been used. These measure the change in amplitude during reflection or transmission of an oscillating system at values greater than those of the systems described earlier by several orders of magnitude. Examples of these devices are the resonant wire method [199] or oscillating quartz structures. Most sensors are

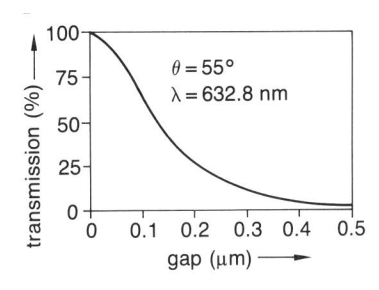

Figure 7.35 Transmission as a function of the size x of the air gap.

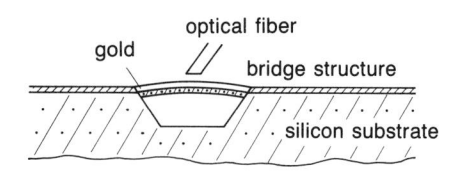

Figure 7.36 Optically excited micromechanical resonant structure [165].

stimulated electrically by means of coils or electrodes. In this way some of the advantages of optical fibers are squandered. More elegant is the direct optical excitation of the mechanical system using, for example, photoacoustic or photothermal methods. Thus photothermal excitation of micromechanical structures and the fiber optic sensing of their resonance form the basis of a completely new family of frequency-coded FOSs which are easy to evaluate and which are path-neutral and network-compatible. The aim is to develop sensors whose resonant frequency is continuously transmitted as a function of the measurand. Figure 7.36 illustrates such a structure which is produced using micromechanical techniques [165]. The lateral bridge dimensions are of the order of a few tens of micrometres with a thickness of a few micrometres. The gold film on the thin 'silicon chord' has two functions. It absorbs a small proportion of the incident radiation, approximately 1%, allows approximately 1% through the Au/Si layer, and reflects 98%. The absorbed energy causes a small temperature increase in the centre of the chord and causes it to bend as a result of the different temperature coefficients of gold and silicon. If the light intensity is frequency-modulated and if this frequency corresponds to the resonant frequency of the chord, then maximum displacement of the reflected light is observed. The evaluation electronics must contain a frequency modulator and, as a result, are far from simple. This type of sensor can operate with either monomode or multimode fibers. Special layers can be applied to the resonant structure in order to produce pressure, temperature and chemical sensors. The development of this type of sensor is still in its infancy [188, 200, 201]. Its great advantage lies in the combination of fiber optic and silicon technologies.

7.4 New fibers

Advances in the development of special fibers for sensor technology can take many forms. The selection of suitable core and cladding materials makes it possible to adjust the numeric aperture, mode distribution and fiber coupling in multimode fibers. The cladding materials can be gaseous, liquid or solid, and a large number of modifications are possible. Special geometries are also of interest. For example, fibers with a *D*-shaped cross-section are particularly well suited for use in distributed pressure sensors [188].

Interesting developments have taken place in the field of monomode fibers [181]. These include the development of fibers with negligible birefringence. The manufacturing process used to produce these fibers minimizes asymmetrical stresses by forming the core and cladding from materials having the same coefficient of thermal expansion. This type of fiber is important in the measurement of magnetic fields. Another fiber type comprises fibers with highly linear birefringence. Interferometric sensors require a stable, linear polarization state which is independent of external excitation. This is achieved by reducing the coupling between the two orthogonal modes. This requirement can be met by generating a high level of linear birefringence in the fiber. One way of realizing this technically is to introduce asymmetrical stresses into the fiber core (Figure 7.37). It can be shown that the best effect is obtained when the sectors which produce the stresses in the fiber have the shape of a bow tie [202]. An alternative method would be to structure the core in such a way that the distributions of the refractive indices are different in the two directions [203].

Fibers with high linear birefringence can operate in two completely different ways. In the first, the orthogonal modes exhibit low transmission losses. They propagate themselves with approximately uniform attenuation. If the same quantity of light is supplied at the input in each mode, then the polarization state along the fiber will change periodically from linear to circular and back again. The beat length l is given by

$$l = 2\pi/(\beta_1 - \beta_2) = \lambda/B \qquad (7.26)$$

with the standardized birefringence

$$B = n_1 - n_2 = \{\lambda/2\pi\}(\beta_1 - \beta_2) \qquad (7.27)$$

where β is the phase constant of the two modes.

If, on the other hand, only one mode is supplied at the input, this is linearly polarized along the entire fiber. The great difference in the phase constants of the two modes considerably reduces the coupling between them. Only if high levels of

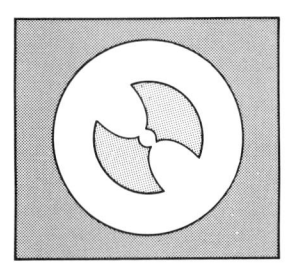

Figure 7.37 Cross-section through a fiber with a high level of double refraction and a 'bow tie' structure [181].

external interference are present is some of the original light coupled into the orthogonal mode. The intensity of this mode is then a measurement of the effect under observation, that is, the external interference. Fibers of this type, in which the attenuation of the two modes is equal but in which the phase constants differ greatly, are known as *polarization-maintaining* fibers.

The second operating mode for fibers with highly linear birefringence is characterized by the fact that one of the two modes is intentionally greatly attenuated. This can be achieved by forming the fiber into a coil which results in high bending losses in one of the modes. This type of fiber is known as a *polarizing* fiber as only linearly polarized light is produced, irrespective of the polarization state at the input. Circular birefringence fibers are very important for the measurement of electric and magnetic fields. They can be produced by mechanically twisting conventional fibers. However, it is more efficient to produce fibers from preforms, in which case the core does not run along the length of the longitudinal fiber axis but instead follows a helical path.

Elliptical birefringence fibers can be manufactured by twisting a fiber with a high level of linear birefringence during crystal growth. This type of fiber is primarily suited to the detection of Faraday rotation.

8 Chemical sensors

Chemical sensors record the concentration of particular particles (atoms, molecules or ions in gases or liquids) using an electric signal. In cases concerned with the detection of specific biological substances, the devices involved are known as *biological* sensors. These are often treated as a separate class of chemical sensor.

Chemical sensors are very different from physical sensors. In the first place, the number of chemical species which act on the sensor is usually very high. Recall that approximately 100 physical measurands can be recorded using physical sensors. In the case of chemical sensors, this number is larger by several orders of magnitude. An example of this is the number of compounds for which tests are carried out in medical laboratories. Secondly, the chemical sensor must be 'open' to the medium which is to be measured and cannot be packaged off as in the case of temperature sensors. This means that it is exposed to undesirable effects such as light or corrosion.

The detection of particular particles as defined above is in many cases no real problem since it is a task that can be performed using the techniques of analytical chemistry, for example with the aid of mass spectrometers, gas chromatographers, or optical or magnetic procedures. In comparison to these procedures, which generally require costly equipment and function discontinuously, there is a pressing need for chemical sensors with properties such as:

- small, robust and reliable construction;
- microelectronics compatibility;
- reproducibility;
- selective and rapid response;
- greatest possible independence from environmental parameters;
- manufacturability using conventional microelectronic methods.

These requirements mean that there is an extensive range of applications. Examples are the measurement of emissions and environmental protection, measurements of imissions, prevention of fire and explosions, process measurements (for example in chemistry, the food industry, biotechnology), medicine, automobile technology, household equipment, water preparation and sewage analysis, surface and materials analysis.

116   Chemical sensors

The ideal of researchers in the field of chemical sensor development is to accommodate a complete, integrated analytical chemistry laboratory on a single chip. However, in reality this is still a very remote prospect.

Millions of chemical sensors have been manufactured. The most widespread of these is the Taguchi SnO_2 sensor for the detection of reducing gases, followed by O_2 sensors which are based on ZrO_2 ion conducting sensors. This development has taken place in spite of the fact that the mechanisms underpinning the functioning of chemical sensors are often not fully understood. Today, ions in solids, membranes or junctions often determine the operation of commercially available sensors.

In recent years the need for chemical sensors has increased greatly. Reasons for this are the increasing complexity of many processing procedures, as well as the desire to economize in the use of energy and raw materials and to reduce environmental pollution. The upshot of all this is that many researchers are now devoting energy to the development of chemical sensors with specific properties for particular applications and with known operating principles. Research and development work is proceeding in two stages. One is the starting point for the development of new sensors which are empirically optimized for use under realistic application conditions and which are characterized by their sensitivity, cross-sensitivity, long-term drift and ageing conditions. The measurand can be direct current conductivity, complex impedances, capacitance, or potentiometric and amperometric quantities. These phenomenological parameters which formally characterize the sensor are then correlated with the atomic construction of the sensor. It is this that makes systematic modification and optimization of the sensor possible in the second stage. Of course, this presupposes the existence of the corresponding physical investigative techniques and theoretical background. The discussion of both will be left until later. In the 1980s Göpel and colleagues provided some stimulating ideas for research into the development of chemical sensors [204–206]. For other overviews refer to [207–213].

8.1 Detection principles and chemical sensor requirements

A simple sensor for the detection of specific molecules from a molecular mixture is schematically represented in Figure 8.1. In this example the linear triatomic molecule CO_2 forms an adsorption complex in the gas with the atoms at the surface of the sensor. This complex can either accept electrons from or donate them to the conductance band of the sensor. The properties of the molecule which allow it to be an acceptor or donor at the surface cause the concentration of free carriers in the semiconductor to decrease or increase. This brings about a corresponding change in conductivity which makes the electronic detection of the particle possible. In principle very different sensor properties G can be used for detection in the

Detection principles 117

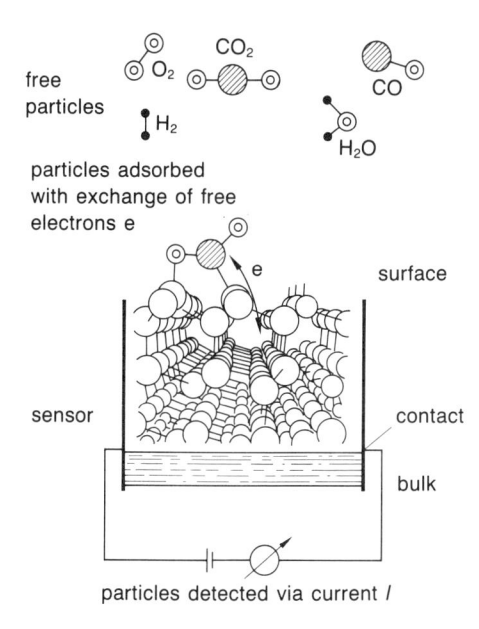

Figure 8.1 A semiconductor gas sensor [204].

selective interaction of particles with the sensor. Table 8.1 lists some of these. An ideal chemical sensor reacts specifically to only one particle type with the sensor property G listed in Table 8.1. In reality such a high degree of selectivity cannot be obtained. Cross-sensitivities mean that the sensor also reacts to other particles. A gas sensor should at least exhibit a reversible response to changes in partial pressure and temperature. This would make the sensor property G a partition function in the thermodynamic sense [206, 214], as indicated in Table 8.2, point 1. Table 8.2 also illustrates that a sensor signal is thermodynamically possible if the partition function G, that is to say, the differential value of the free enthalpy, is negative for the interaction between the detected particles and the sensor surface. Point 3 makes it clear that adsorption effects are preferred for particle detection at low temperatures. In contrast, sensitivity increases with temperature when particles are detected using

Table 8.1 Characteristic sensor properties used for detection in chemical sensors [331]

Liquid electrolyte sensors:	electric potential, current, capacitance
Solid electrolyte sensors:	electric potential, current
Conductivity sensors:	electrical conductivity
Field effect sensors:	potentials
Calorimetric sensors:	adsorption or reaction heat
Photochemical sensors:	optical dimensions as a function of frequency
Mass-sensitive sensors:	adsorbed particle mass

Table 8.2 Thermodynamic aspects of chemical sensors [331]

Overview (1) of the preconditions for reproducible sensor signals; (2) of partition functions and velocities of specific sensor–particle interactions; (3) of the phenomenological thermodynamics of general sensor–gas interactions; and (4) of the relationship between the degree of coverage θ, free enthalpy G and spectroscopic data ε_i where the energy states ε_i characterize the atomic structure of the sensor surfaces and the boundary layers.

1. Requirement: sensor property G is a partition function

Objective

$$dG^* = \left(\frac{\partial G^*}{\partial p_1}\right)_{p_i \neq 1, T} dp_1 + \left(\frac{\partial G^*}{\partial p_2}\right)_{p_i \neq 2, T} dp_2 + \cdots + \left(\frac{\partial G^*}{\partial T}\right)_{p_i} dT$$

where

$$\int dG^* = 0$$

2. Thermodynamic and kinetic aspects of chemical and biochemical sensors
   (a) Why are there sensor–particle reactions?
   (b) How fast are these sensor–particle reactions?

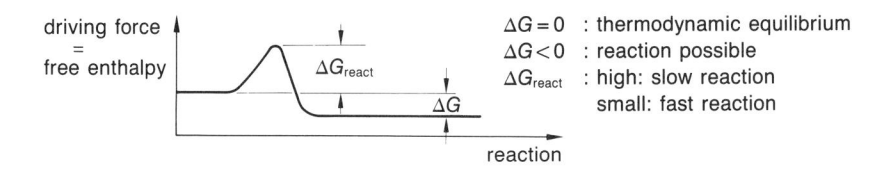

$\Delta G = 0$: thermodynamic equilibrium
$\Delta G < 0$: reaction possible
ΔG_{react} : high: slow reaction
         small: fast reaction

3. Why free enthalpy G and not 'energy'?

Thermodynamics: $\Delta G = \Delta H - T\Delta S = \Delta(U + pV) - T\Delta S$
Examples:

(a) Chemisorption $\Delta U(\Delta H) < 0$ and $\Delta S < 0$
   → good results at $T = 0$ K
   desorption when $|T\Delta S| > |\Delta H|$

(b) Point defects $\Delta U(\Delta H) > 0$ and $\Delta S > 0$
   → negligible at $T = 0$ K
   improved at high temperatures if $|T\Delta S| > |\Delta H|$

4. Relationship between G and the energies ε_i of electrons, photons, plasmons, ... -ons (i.e., elementary stationary states of the material)?

Statistical thermodynamics:

$$\varepsilon_i \leftrightarrow Q = \Sigma\, e^{\varepsilon_i/kT} \leftrightarrow G = -kT\left(\ln Q - \frac{\partial \ln Q}{\partial \ln V}\right) \leftrightarrow \Theta = f(P, T), \text{ etc.}$$

the point defects of the sensor. As most sensors operate at an intermediate temperature range, both effects must be reckoned with. At the same time as the desired reaction, a series of other thermodynamically or kinetically determined reactions is generally taking place at the sensor surface. These can lead to irreversible changes and thus to long-term drift, although this can be avoided by selecting a suitable sensor material. Modern methods of boundary layer analysis make it possible to register information concerning atomic configurations, area coverage, impurities and bond strength both before and after the interaction of the sensor with the particles. One of the items of information provided by these data is the elementary excitation energies of the various stationary states of the system. This establishes a relationship with the state sum Q as specified in Point 4 of Table 8.2 and thus with phenomenological observations. Given Q, adsorption isotherms can be used to provide information about, for example, the degree of coverage $\theta' = f(P')_{T=\text{const}}$ of molecules adsorbed at the surface of a gas sensor as a function of the partial pressure and the temperature. This principle can be applied to other parameters.

In summary, we can say that the use of the most varied analytical methods is absolutely necessary for the selective development of chemical sensors.

8.2 Design types

Chemical sensors can be classified in accordance with their varying fields of application, the particles they are intended to detect or their physicochemical operating principle. A distinction is also frequently drawn between chemically sensitive semiconductor components and others [215, 216]. In this text I shall use the following categorizations

1. *Conductivity sensors.* In these sensors the interaction of the gas with the solid (semiconducting metal oxide or organic semiconductor) causes a change in conductivity. A change in resistance can also be caused by a change in the temperature of the sensor material.
2. *Structured semiconductor sensors.* These are modified semiconductor components in which changes in electric double layers at phase boundaries are used for measurements.
3. *Electrochemical sensors.* These sensors use the catalytic effect of specific electrodes for gas detection or the selective interaction of molecules or ions with fixed membranes in fluid systems.
4. *Solid electrolyte sensors.* These employ ionic conduction at negligible electron conduction levels.
5. *Chemically sensitive FETs.* In these devices the interaction of ions or molecules at an ion-selective or gas-sensitive layer located in the gate region of an FET causes the gate potential of the transistor to change. In principle,

this type of sensor is a combination of a potentiometric sensor (the sensing layer) and a charge amplifier.
6. *Other types of sensor.* These include sensors which react indirectly to chemical processes, such as optical sensors, optothermal sensors, quartz microbalances, SAW sensors and biological sensors which are used especially to detect biological substances.

8.2.1 Conductivity sensors

Because of their chemical and thermal stability it is metal oxides that are predominantly used in practical applications. At temperatures of approximately 500°C their electrical resistance exhibits a high degree of dependence on the concentrations of certain gases. At temperatures of 200–500°C metal oxides with n-type electron carrier paths, such as SnO_2, ZnO or Fe_2O_3, primarily respond to oxidizable gases such as H_2, CH_4, CO, C_2H_5 or H_2S and increase their conductivity. In contrast, p-type semiconductors such as CuO, NiO or CoO respond primarily to reducible gases such as O_2, NO_2, Cl_2. Alongside these binary metal oxides, ternary or quaternary metal oxides and compounds of these are used in homogeneous gas sensors. Metal oxide sensors are frequently modified and optimized through the addition of metal atoms such as Pd, Pt, Cu, Au and Ag [214, 217, 218]. This gives the sensors a certain degree of selectivity. Table 8.3 provides an overview of the metal oxides used for particular gas sensors. For technical reasons, metal oxide sensors based on SnO_2, ZnO_2 or Fe_2O_3 have the greatest significance in practical applications.

The first homogeneous gas sensors were polycrystalline sensors. The best known of these is the Taguchi gas sensor (TGS) which is based on SnO_2 and is produced in large quantities by the Japanese company Figaro. This sensor is installed as a fire alarm in many Japanese homes. Figures 8.2(a) and 8.2(b) display a number of possible design variants, while Figure 8.2(c) depicts a commercial Taguchi sensor. Detection of reducing gases is performed using the conductance value G with response times of the order of minutes. The current limit of detection using SnO_2 sensors is 0.2 ppm. Long-term drift, primarily in the form of diffusion and sintering effects at grain boundaries, is unavoidable. Basically, G is related to the concentration c_i or the partial pressure p_i of the gas to be detected by

$$G \sim c_i^{n_i} \tag{8.1}$$

where n_i is an empirical constant which is positive and smaller than 1 in the case of reducing gases. However, this sensor is not well suited to making quantitative measurements. For example, in the absence of reducing gases the conductance value is dependent on the partial pressure of oxygen and water. Moreover, although a large number of reducing gases can be detected, there is always a certain degree of cross-sensitivity. Cross-sensitivity means that equivalent changes in resistance can be caused by changes in the concentrations of different gases. There are a number of

Table 8.3 Selected metal oxides for particular gas sensors

Oxides in the corresponding semiconductor sensor	Gases detected
TiO_2, Fe_2O_3, CoO, ZnO, ZrO_2, SnO_2, La_2O_3	O_2
Cr_2O_3, NiO, ZnO, ZrO_2, SnO_2, In_2O_3	CO
Fe_2O_3, Fe_3O_4, Co_3O_4, ZnO	CH_4
SnO_2, VO	NO_x
ZnO, Al_2O_3, SnO_2	halogens

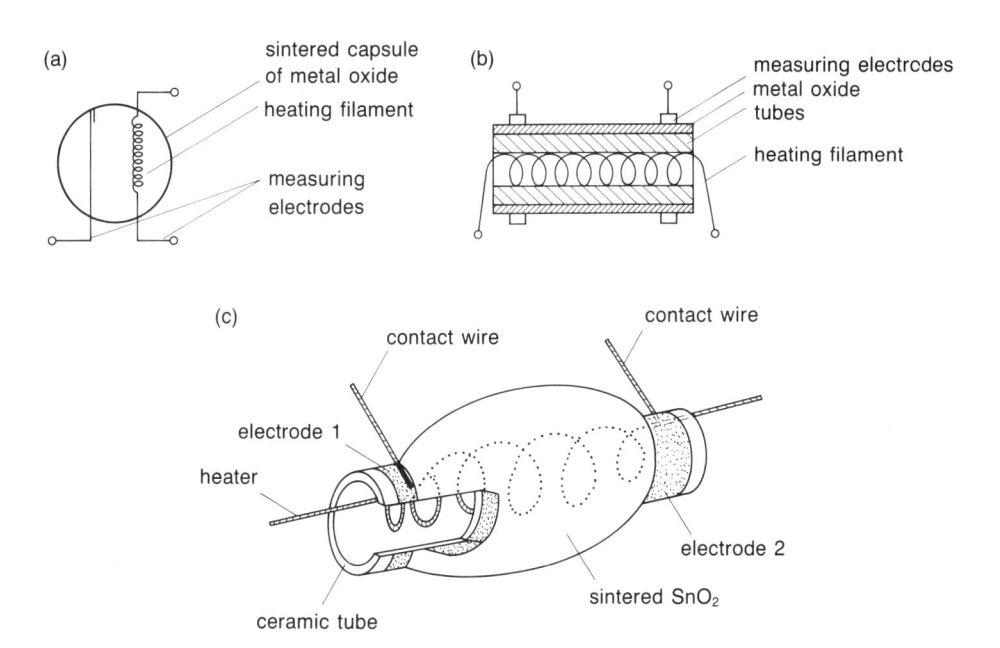

Figure 8.2 Homogeneous semiconductor sensors [204]: (a) polycrystalline sensor with sintered platinum heating filament; (b) polycrystalline sensor with separate heating filament; (c) commercial Taguchi sensor.

ways of modelling this cross-sensitivity. For example, an attempt has been made to describe it formally by the following parameterization [219]:

$$(G/G_0)^\alpha = \left(1 + \sum_{j=1}^{N} k_j c_{1j}^{n_{1j}} c_{2j}^{n_{2j}} \ldots \right) / c_{O_2} \qquad (8.2)$$

Equation (8.2) describes the change in the conductance value on detection of N gases with concentrations c_{ij} using N summands with α, k_j and n_{ij} as the empirical parameters. G_0 is the conductance value under standard conditions. Although this equation can often be simplified, it still has to be modified for certain gases. The analytical form required determines, for example, whether the sensor can be used for gas detection or if pattern recognition in a sensor array should be used.

The behaviour of the conductance value in n-type semiconductors can be explained in qualitative terms by the competing influence of negative adsorbed oxygen and the associated formation of electron-depleted surface layers at the surface or between the crystallites. Reducing gases decrease the amount of adsorbed O_2 and thus cause a smaller extension of the electron-depleted surface layers and, in consequence, an increase in conductivity [224]. In principle this conductivity effect can also be caused by the reaction of the reducing gas with the lattice oxygen. This causes defects which can diffuse into the bulk at high temperatures. The measurand, namely overall conductance, is determined by the proportions of electrons and ions at the surface, at grain boundaries and in the bulk of the sensor. A great deal of effort is being devoted to providing the quantitative underpinning or extension of this theoretical explanation; see, for example, [220–223].

Recently thin-film and single-crystal structures have been developed as homogeneous gas sensors (Figure 8.3). In comparison to polycrystalline material, thin films have low concentrations of grain boundaries but still possess a good ratio of surface to bulk atoms. The most frequently used materials are ZnO, SnO_2 and TiO_2 [218, 225, 226, 229]. These sensors have a high sensitivity and possess response times of the order of minutes. However, they exhibit a relatively high degree of cross-sensitivity. Satisfactory long-term stability and reproducibility have not yet been achieved. It is hoped that the use of epitaxial layers will bring about an advance in these areas. So far only ZnO single crystals have been used for sensors. However, these sensors are very expensive to produce.

Thick-film technology is also of interest for metal oxide sensors since this permits a greater degree of reproducibility than can be achieved using sintering techniques. Despite this, only a few examples are known [224]. An interesting example of a thick-film sensor is a multisensor for odour detection, as illustrated in Figure 8.4 [241].

The principal areas of application of homogeneous semiconductor sensors are in domestically installed gas supplies, warning systems for the detection of CO, for the detection of H_2S in refineries, and for testing the level of alcohol in the breath. Semiconductor sensors which detect changes in surface resistance are also used to

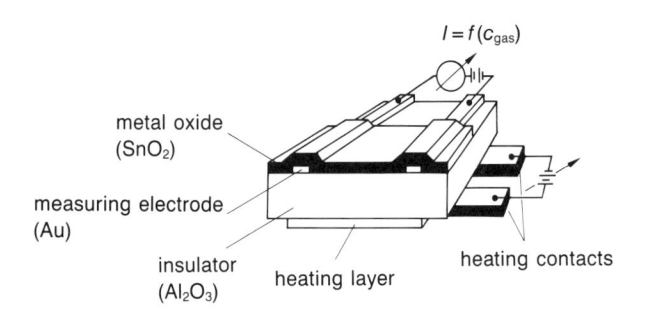

Figure 8.3 Thin-film metal oxide sensor [204].

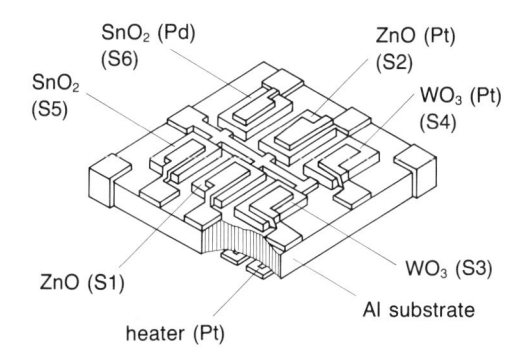

Figure 8.4 Multisensor using thick-film technology [241].

measure relative humidity. The sensor materials employed for this are $TiO_2-V_2O_5$ [227] or $MgCr_2O_4-TiO_2$ [228]. Some of these sensors possess a very rapid response time of the order of seconds.

Sensors made from organic semiconductors also experience a fall in resistance when they are exposed to reducing or oxidizing gases. Although this has long been known [230, 231] and despite the fact that they can operate at approximately 150°C, that is to say, at much lower temperatures than metal oxide sensors, they are still employed in relatively few applications. This is in part a result of the fact that questions of purity or selective impurities, the structural make-up of the sensitive material, as well as contact-related questions are decisive for the sensitivity, stability and reproducibility of the sensor. It follows that the properties and the conditions for the manufacture of these materials must be understood in considerable detail. This class of materials includes metal-phtalocyanine complexes and semiconducting polymers such as polypyrol. A polycrystalline film of PbPc, which is heated to 150°C, is used for the selective detection of nitrogen oxides of the type NO_2 [232,233] (Figure 8.5). The detection limit is in the ppb-range. Such a sensor is very suitable for the measurement of toxic gases in the environment such as NO_2/N_2O_4 mixtures. CuPc is a suitable thin film for the detection of NH_3 [234]. H_2Pc and CoPc are also being examined as potential sensor materials [234, 235].

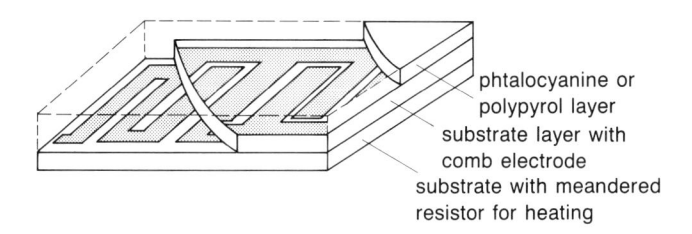

Figure 8.5 Gas sensor based on a phtalocyanine or polypyrol layer [232].

124 Chemical sensors

Interesting avenues of development have been opened up for the use of semiconducting or conducting polymer materials in sensor technology. For example, at 150°C polypyrol displays a sensitive resistance change in the presence of NH_3 [236]. The combination of these organic semiconductors with inorganic sensor materials is an intriguing prospect for the future [237].

As a final example of a conductivity sensor let us take the catalytic gas sensor which has now been in use for many years. Inert sintered bodies, consisting primarily of ThO_2 or γ-Al_2O_2, with an internal filament and whose surface is covered by a catalyst, react with reducible gases at approximately 550°C (Figure 8.6). This surface reaction generates heat which is determined by measuring the resistance of either the pellistor or the Pt wire. These devices have recently been the focus of much attention since they can function as selective sensors when a suitable catalyst is chosen.

8.2.2 Structured semiconductor sensors

In these devices the change in the electric double layer at an insulator/metal phase boundary is used to make measurements [238, 239]. They are of particular interest for the detection of H_2 at room temperature. The metal used is normally palladium. This has two important properties. First, it is catalytically active. It dissociates H_2 molecules to form hydrogen atoms at its surface. Second, it has a high level of solubility for H atoms. If a Pd-MOS diode (Figure 8.7) or a Pd-MOS transistor, known as a GasFET, in which the metal gate is made of Pd [240], is exposed to hydrogen, the latter is dissociated at the external metal surface. The hydrogen atoms then diffuse through the metal and are adsorbed at the metal–insulator interface. A dipole boundary barrier is formed, and this reduces the barrier which is generated if there is no H_2 supply. This can be detected by changes in the characteristic curves, for example in the capacitance–voltage curve in a MOS diode, or by the shift in the threshold voltage in a transistor.

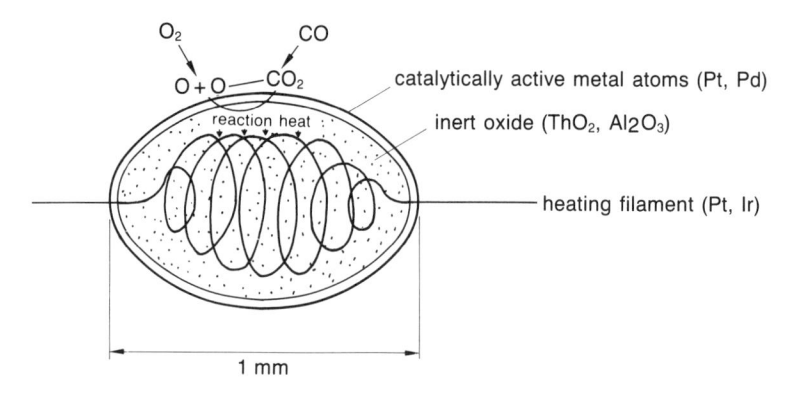

Figure 8.6 Structure of a catalytic gas sensor.

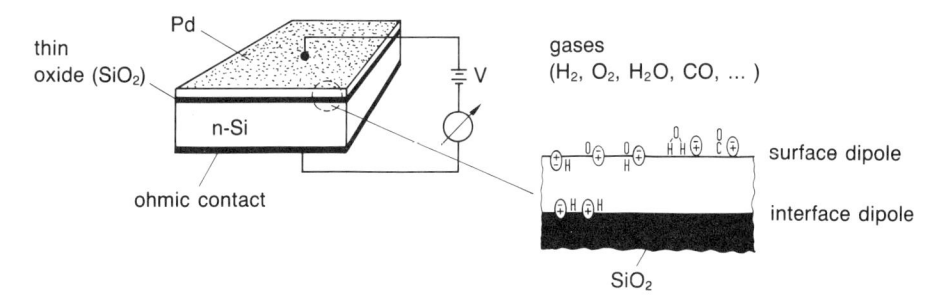

Figure 8.7 Structure of a Pd-MOS diode [204].

Attempts have also been made to detect other gases such as CO using this technique. However, the majority of these sensors exhibit a high degree of cross-sensitivity in the presence of H_2. For this reason the MOS structures are modified by providing openings in the gate metal. The gases can pass through these openings directly to the metal oxide interface where they shift the work function of the metal. Another variant makes use of gap–gate transistor structures (Figure 8.8) [242, 243]. The special short channel transistor structure disables the diffusion of H_2 in the gate metal and thus reduces the cross-sensitivity to H_2.

An alternative solution is to integrate an H_2 sensor with a CO_2 sensor on a single chip and to use the current value of the H_2 concentration to correct the signal from the CO_2 sensor during signal preparation (Figure 8.9). In fact this represents a pattern recognition procedure. A new development with an extended lifetime consists of FETs with a suspended microgrid. These are also known as suspended gate CHEMFETs [243]. In these devices the gate metal is preceded by an additional space which, in the case of GasFETs, is permeable to gases (Figure 8.10). The suspended grid above the gate insulator is made of Pt or Au. Applying a Pd layer to this creates an H_2 sensor. If a conductive polymer layer, such as polypyrol, is deposited on the metal grid, then the sensor is highly sensitive to alcohols. In both cases the reaction of the gas with the surface of the suspended metal grid or with the surface of the insulator causes a change in the electric field which is detected in the modified drain current.

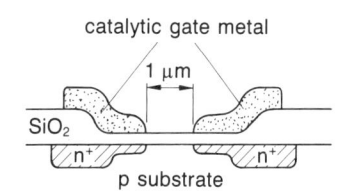

Figure 8.8 Illustration of a GasFET with a gap–gate structure [331].

126 Chemical sensors

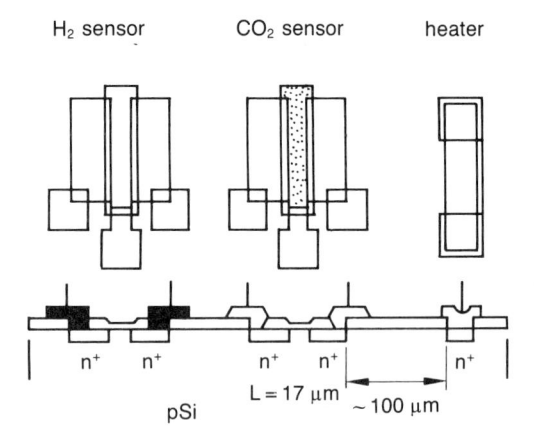

Figure 8.9 Integration of CO_2 and H_2 sensors and a heater on a silicon sensor [331].

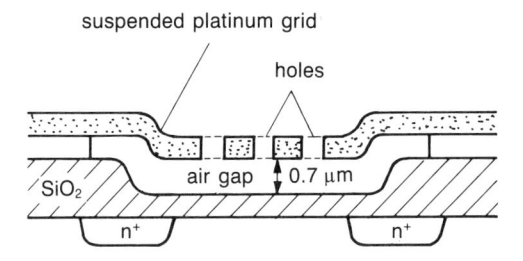

Figure 8.10 GasFET with suspended Pt grid above the gate [331].

8.2.3 Electrochemical sensors

Substances in gases and liquids can be detected electrochemically if they enter into reactions in the electrochemical cell or if they influence reactions taking place in the cell [244]. An electrochemical cell consists of an electrolyte, a measuring electrode, a counterelectrode and a reference electrode. Normally in the cell a substance S_{red} is anodically oxidized to S_{ox} or a substance S_{ox} is cathodically reduced to S_{red}.

$$S_{red} \leftrightarrows S_{ox} + ne^- \tag{8.3}$$

where n is the number of electrons required, S_{red} is the reduced form of the substance and S_{ox} is the oxidized form. In these reactions, electrons are either consumed or released. At equilibrium these electrons cause a corresponding potential to appear at the electrode or, if the material is converted, a current appears.

Potential generation or current flow are used as the detection methods in electrochemical sensors. The method of measurement of voltages or currents is known variously as *potentiometric* or, depending on the preset potential, as *amperometric*, *polarographic* or *voltammetric*. The operating principles of the sensors are categorized in accordance with these methods.

Figure 8.11 presents a schematic illustration of a potentiometric sensor. At equilibrium the reaction which determines the potential at the electrodes is described by the Nernst equation

$$\varphi = \varphi_0 + (RT/nF) \ln a_{ox}/a_{red} \tag{8.4}$$

where φ is the potential of the measuring electrode, φ_0 is the standard potential, R is the gas constant, T is the absolute temperature, F is the Faraday constant, n is the reaction velocity, and a designates the activity of the components S_{red} and S_{ox} which are related to the concentration c and the partial pressure p. At high temperatures in particular there are deviations between a_i and c_i/c_0. The potential of the measuring electrode is measured as a voltage against the potential of a reference electrode, for example a normal hydrogen electrode or the calomel electrode. The use of potential measurement methods is particularly worthwhile when only one reaction progresses reversibly and there are therefore no mixed potentials. The reaction velocity should be fast. This group is represented by the ion-sensitive electrodes, the best-known of which is the long-serving glass electrode for determining pH values [245]. It is used to detect H_3O^+ ions. Figure 8.12 shows a pH-measuring cascade with a pH glass electrode and reference electrode. Ion exchange reactions at the glass electrode cause a jump in potential at the electrode surface which is in contact with the measured solution, the size of which is dependent on the H_3O^+ concentration. The inside of the glass electrode undergoes a constant jump in potential in comparison to the reference electrolyte. The Ag/AgCl electrode is used to tap the voltage. The potential is measured relative to the reference electrode. Ideally the potential obeys the Nernst equation. The voltage is logarithmically dependent on the concentration of H_3O^+ ions.

It is also possible to modify the membrane of the glass electrode in order to obtain electrodes for other ions, for example Na^+, K^+, F^-, etc. This is done by using materials with selective ion exchange potentials, selective ion conduction or selective ion inclusion. Figure 8.13 illustrates the principle underlying the construction of an

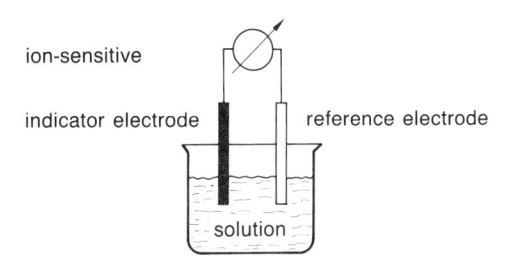

Figure 8.11 Potentiometric sensor.

128   Chemical sensors

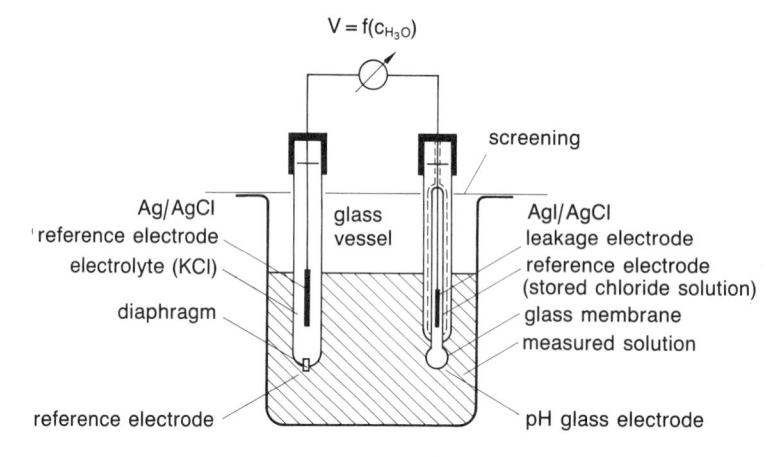

Figure 8.12   A pH-measuring cascade with glass and reference electrodes [204].

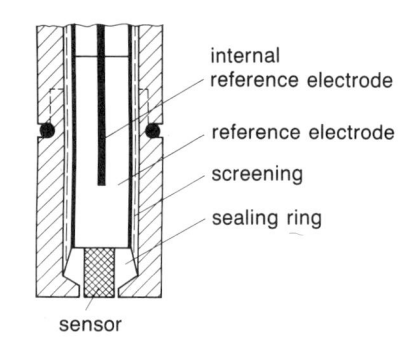

Figure 8.13   Ion-selective electrode with a solid membrane [204].

ion-selective electrode with a solid membrane. An example membrane is LaF_3 which acts as a conductor of F^- ions. Other ions can be detected by using different metallic salt membranes [246].

Another large group of ion-selective electrodes possesses liquid or polymer membranes with built-in ion exchangers or neutral ion carriers. The liquid can be present either in the form of a gel or as a cellulose acetate membrane. The most frequently used polymers are PVC, polyethylene and silicon rubber. Valinomycine, embedded in PVC plastisol, is an example of a neutral ion carrier which is often used in K^+-selective electrodes.

Ion-selective electrodes can also often be used as gas sensors, for example for the detection of soluble gases such as NH_3, CO_2 and SO_2. If the electrochemical detection reaction is preceded by another reaction, for example an enzyme reaction which generates or consumes species which contribute to the detection process, then these can also be used for biological sensors. Using ion-sensitive electrodes it is possible to achieve relative error levels of 1%. The response time in the 1 ppm range is approximately 1 minute for 90% values. The disadvantage of these electrodes is that complex electrometer circuits with extremely high input resistances are often required. In addition, they are relatively large (typical dimensions: length 100–150 mm, diameter 10 mm) and are therefore mechanically susceptible to errors. These disadvantages can be avoided in ion-sensitive FETs.

Ion-selective electrodes are used in a great number of process measurement applications, in clinical analysis and research. Example applications are the measurement of ions in blood, urine, seawater, sewage, domestic water, foodstuffs, and consumer goods [247].

If more than one reaction is proceeding then the mixed potential that is measured at the electrodes cannot be attributed definitively to the component which is being analyzed. In such cases, current measurement should be preferred to potential measurement. When current measurement techniques are employed the potential at equilibrium of the measuring electrode is intentionally held at either high or low potential values. This means that a cathodic or anodic current flows through the measuring electrode and chemical transformation takes place. The substance which is to be detected is electrochemically reduced or oxidized. The measured signal is the Faraday current I:

$$I \approx F(dN_i/dt) \tag{8.5}$$

where dN_i/dt represents the conversion of component i in unit time. This current is dependent on the potential of the electrodes as well as on the concentration of the component to be measured. If an unambiguous relationship between current and concentration is to be obtained, then care must be taken that the component which is to be detected is fully transformed at the measuring electrode. In this way, potential independence is achieved. The simplest way of doing this is to use a diffusion barrier. The potential of the electrode is removed so far from a state of equilibrium that all the diffused molecules or ions are converted at the electrode. This means that the concentration at the electrode surface of the species which is to be measured is zero. The diffusion limit current is approached. The simple relationship

$$I_{\lim,i} = \{(nFAD)/\delta\}c_i \tag{8.6}$$

applies, where δ is the thickness of the diffusion layer, D is the diffusion coefficient, A is the surface area of the electrode and c_i is the concentration of the component i. The current–voltage curves shown in Figure 8.14 are obtained. There are two ways of setting the potential of the measuring electrode in the region of the diffusion limit current:

- selection of a reference electrode with a suitable electrode potential;
- electronically, using a potentiostatic circuit.

130 Chemical sensors

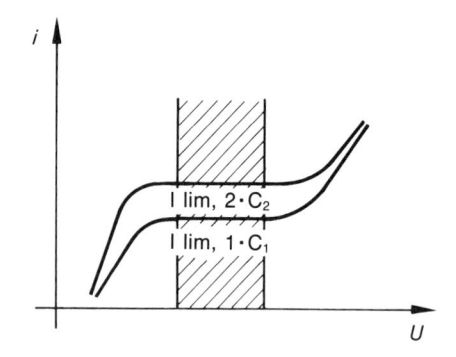

Figure 8.14 Current–voltage curves with a diffusion-regulated limiting current range (shaded) [244].

Cells containing two or three electrodes are used for measuring currents. Common to all these is the fact that the gas which is to be measured can only reach the catalytically active measuring electrode by crossing a diffusion barrier. Figure 8.15 illustrates a dual-electrode cell used for detecting SO_2. A constant voltage between the measuring electrode and the counterelectrode is preset. The counterelectrode may only become slightly polarized by current flow so that the potential of the measuring electrode can remain constant. This problem is bypassed in the three-electrode cell in which the potential is maintained at a predefined value by the potentiostatic control and the constant-potential reference electrode (Figure 8.16). Different anode and cathode materials are used depending on the gas which is to be detected (Table 8.4). Teflon (PTFE) is often used as the membrane.

The Clark cell [249] is an important O_2 sensor for both industrial and medical applications [248, 250]. The electrochemical two-electrode cell is separated from the environment by an O_2-permeable, hydrophobic membrane (Figure 8.17). A voltage

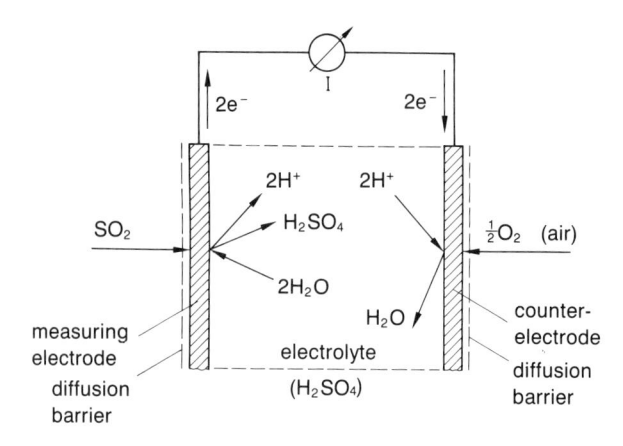

Figure 8.15 Dual-electrode cell (diagrammatic) for the detection of SO_2 in air [244].

Table 8.4 Examples of chemical cells used in gas detection

Gas component	Anode	Anode reaction	Cathode	Cathode reaction
H_2S	Ag	$H_2S + 2AG^+ \rightarrow Ag_2S + 2H^+$ $2AG \rightarrow 2AG^+ + 2e^-$	Ag	$\frac{1}{2}O_2 + 2H^+ + 2e^- \rightarrow H_2O$
CO	Pt	$CO + H_2O \rightarrow CO_2 + 2H^+ + 2e^-$	C	$\frac{1}{2}O_2 + 2H^+ + 2e^- \rightarrow H_2O$
Cl_2	Pt	$H_2O \rightarrow \frac{1}{2}O_2 + 2H^+ + 2e^-$	Au	$Cl_2 + 2e^- \rightarrow 2Cl^-$
SO_2	Au	$SO_2 + 2H_2O \rightarrow H_2SO_4 + 2H^- + 2e$	Pt	$\frac{1}{2}O_2 + 2H^+ + 2e^- \rightarrow H_2O$

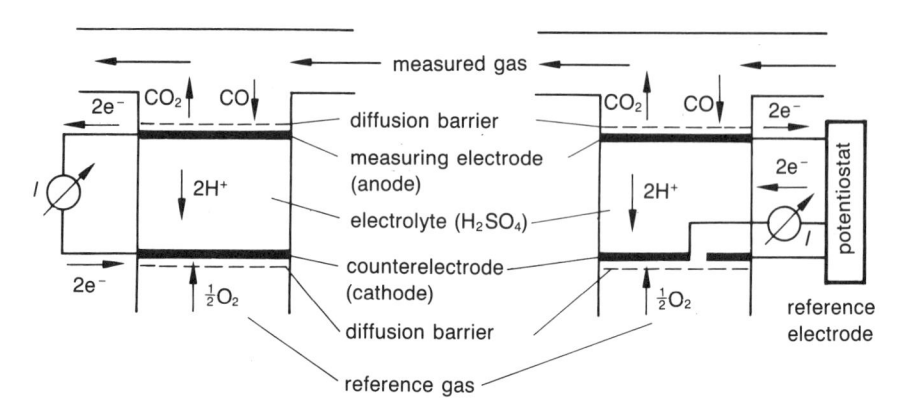

132 Chemical sensors

anode reaction

$$CO + H_2O \rightarrow CO_2 + 2H^+ + 2e^-$$

Figure 8.16 Electrochemical gas sensor for the measurement of CO (potentiometrically regulated three-electrode cell).

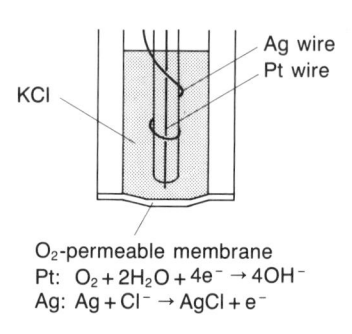

cathode reaction

$$\tfrac{1}{2}O_2 + 2H^+ + 2e^- \rightarrow H_2O$$

Figure 8.17 Clark cell [207].

Pt: $O_2 + 2H_2O + 4e^- \rightarrow 4OH^-$
Ag: $Ag + Cl^- \rightarrow AgCl + e^-$

of 600–800 mV is applied between the electrodes. This causes the sensor to operate in the diffusion limit current range. As the diffusion is temperature-dependent, the temperature has to be measured and its influence compensated for. An interesting variant of the integration of an O_2 and a temperature sensor on one chip is presented in [248] in connection with medical applications. However, the Clark cell is also used for the measurement of oxygen in sewage-treatment plants, in industrial water circulation systems, to minimize corrosive processes or to determine the proportion of dissolved oxygen in surface water.

Amperometric sensors are also used to determine the amount of Cl that is present in drinking water.

8.2.4 Solid electrolyte sensors

Solid electrolyte sensors occupy a special position among the chemical sensors [251]. They make use of the property of certain ionic crystals of transporting electric

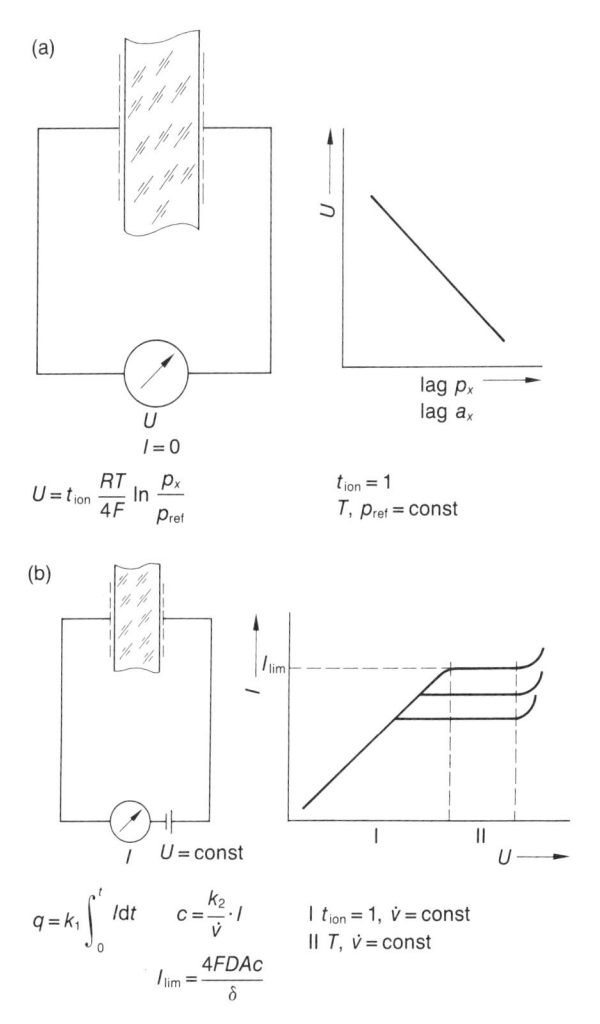

(a)

$$U = t_{ion} \frac{RT}{4F} \ln \frac{p_x}{p_{ref}}$$

$I = 0$

$t_{ion} = 1$
$T, p_{ref} = \text{const}$

(b)

$$q = k_1 \int_0^t I \, dt \qquad c = \frac{k_2}{\dot{v}} \cdot I$$

$$I_{lim} = \frac{4FDAc}{\delta}$$

$U = \text{const}$

I $t_{ion} = 1$, $\dot{v} = \text{const}$
II T, $\dot{v} = \text{const}$

Figure 8.18 (a) Potentiometric solid electrolyte sensor; (b) amperometric solid electrolyte sensor [252].

134   Chemical sensors

current in the form of ions at high temperatures. A distinction is drawn between two basic configurations [252]:

- the zero-current measurement of the potential between a reference and a measuring electrode (Figure 8.18(a));
- the measurement of the ionic current through the application of an external voltage at the electrodes (Figure 8.18(b)).

The operating mode of potentiometric solid electrolyte sensors is based on the Nernst equation, that is to say, if the temperature is constant and the potential at the reference electrode is known, then there is a linear dependence between the cell voltage and the logarithm of the activity or the partial pressure of the component which is to be measured. These sensors are able to cover a wide measuring range. Their rapid display (of the order of seconds) and their broad operating temperature range (100–1800°C, depending on the solid electrolyte material) suggest that potentiometric sensors are particularly well suited to *in-situ* process measurements. The best-known example of a sensor belonging to this category is the O_2 sensor which is based on yttrium-doped ZrO_2 [253, 254]. It is used to measure the O_2 partial pressure in car exhaust fumes and to regulate the fuel–air mix to ensure the most efficient use of catalytic converters. This device is known as a *lambda probe*. Since the mid-1970s more than 10 million lambda probes have been put into operation around the world. The ZrO_2 sensor also plays an important role in the monitoring of oxygen content in combustion gases as well as in other more general gas mixtures, or in the measurement of residual oxygen levels in molten steel and other molten metals or in molten glass. Commercial measuring systems have been produced for these applications. Many of them are used in power stations. By using sensors it is possible to reduce fuel consumption considerably and thus also to cut environmental pollution.

A simplified explanation of the mode of operation of a potentiometric ZrO_2 sensor for the detection of O_2 is presented in Figure 8.19. The porous platinum

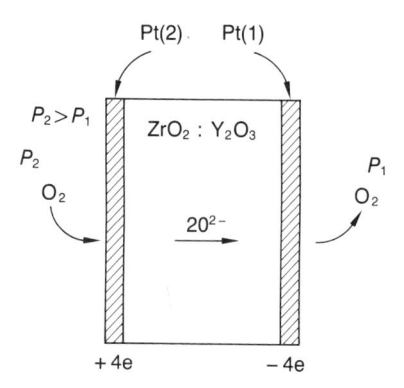

Figure 8.19 ZrO_2 cell [255].

electrode Pt(2) is exposed to a high O_2 content, while Pt(1) is exposed to a low content. At the cathode (Pt(2)) the reaction

$$O_2 + 4e^- \rightarrow 2O^{2-} \tag{8.7}$$

and at the anode (Pt(1)) the reaction

$$2O^{2-} \rightarrow O_2 + 4e^- \tag{8.8}$$

take place. The electron transfer from 2 to 1 causes the build up of an electric field. At equilibrium the electric potential difference is equal and opposite to the chemical potential difference which is caused by the difference in the O_2 partial pressure at the two electrodes. The voltage between the electrodes then has the value

$$U = R(T/4)F \ln(p_1/p_2) \tag{8.9}$$

Figure 8.20 shows the structure of a potentiometric Y_2O_3-doped ZrO_2 sensor or lambda probe. The ceramic is coated on both sides with a porous Pt catalyst and separates the reference and measuring media. To protect against impurities, the electrode which is exposed to the exhaust fumes is covered by an additional porous ceramic layer. The air–fuel ratio, also known as λ, is defined as

$$\lambda = \frac{\text{supplied air volume}}{\text{theoretical air requirement}} \tag{8.10}$$

This ratio gives the sensor its name. λ = 1 corresponds exactly to the stoichiometric mixture of oxygen (air) and fuel (e.g., petrol). In the case of petrol, the ratio for complete combustion of the fuel is approximately 14.6:1. In such a case, ideal

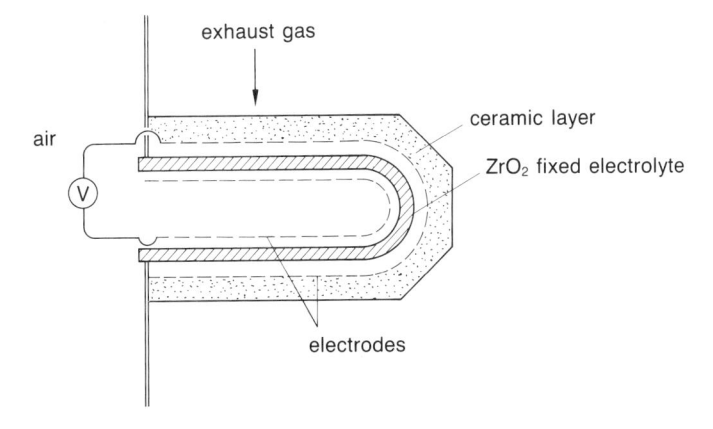

Figure 8.20 Lambda probe for measuring oxygen in exhaust gases.

136 Chemical sensors

combustion has been achieved. The pollutant output from a vehicle equipped with a catalytic converter is at its least when a stoichiometric mixture is present. Passing through the range $\lambda < 1$ (thick mixture) to $\lambda > 1$ (thin mixture) increases the oxygen concentration by more than ten orders of magnitude. This also causes a great change in the potential of the probe (Figure 8.21). This change is very well suited to electronic analysis. This means that precise measurement of the point $\lambda = 1$ is possible. This permits regulation of the fuel–air mixture and guarantees the perfect functioning of three-way catalytic converters.

In the case of vehicles which are operated without a catalytic converter it is sensible to use $\lambda > 1$, that is to say, to thin the fuel mixture. This results in optimum fuel consumption with minimum NO_x emissions [256, 257]. Both potentiometric (with $\lambda = 1.5$) and diffusion-limiting current probes, that is, ZrO_2-based amperometric probes (with $\lambda = 2$), are used. Amperometric solid electrolyte sensors require no reference medium, that is to say, there is no need to separate the electrodes. The ion current can flow in either direction depending on the polarity of the applied voltage. If the sensor is to operate near the diffusion-limiting current then it is necessary to provide diffusion barriers for the component x in the measuring medium at the electrode which is consuming component x, for example by creating a particular geometry (gap) or by introducing a porous layer between the measuring medium and the electrode. Figure 8.22 illustrates this type of limit current probe for λ measurement in vehicle exhaust gases. The limit current is proportional to the

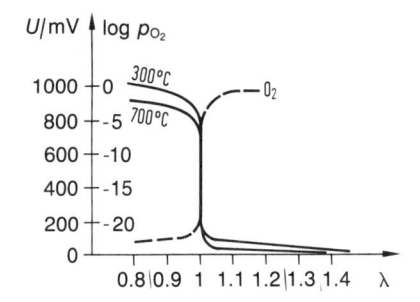

Figure 8.21 Relationship between probe voltage or oxygen content and λ value [255].

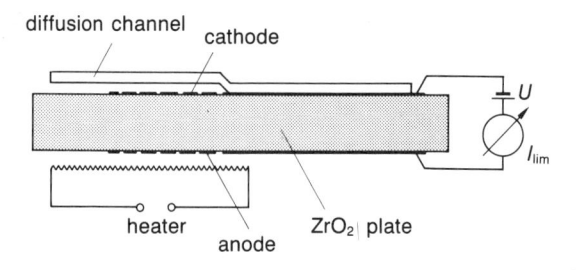

Figure 8.22 Limiting current probe for λ measurements in vehicle exhaust gases [252].

oxygen concentration. Both potentiometric and amperometric sensors can be heated to the temperature of several hundred degrees Celsius which is necessary for ion transport, either through internal heating or by using the heat of the exhaust gases. Of interest is the harnessing of these two sensors in order to cover a wide range of concentrations [255, 258, 259]. Figure 8.23 illustrates the principle. The potentiometric sensor measures the concentration in the chamber. Depending on the direction of current, the current sensor pumps O_2 into or out of the chamber. This effect can now be used to evaluate the jump in potential of the potentiometric O_2 sensor at the stoichiometric ratio in the excess O_2 range ($\lambda > 1$). By varying the pump current it is possible to shift the characteristic curve to higher values of λ. This type of oxygen sensor with integrated oxygen pump has been released using thick-film technology [260]. Other developments for use in thin mixtures are investigating the use of different solid electrolytes such as $SrTiO_3$ [261].

Apart from their use in O_2 detection, ZrO_2 sensors can also be employed for the indirect detection of H_2 and CO or H_2O vapour and CO_2 [262]. The measured quantity is the bound oxygen at chemical equilibrium. Coupling the measuring and pump cells makes it possible to measure the combustible contents of inert gases [263]. Other examples are listed in [264].

The use of ZrO_2 sensors is not limited to the measurement of O_2 in gases. They are also of great interest in connection with measurements in molten substances such as molten steel, molten copper [265] or in liquid Na [252]. Although ZrO_2 is by far the most frequently used conductor of oxygen ions, CeO_2, TiO_2 and Nb_2O_5 are attracting increasing interest [266]. Proton, halogen ion and cation conductors are at the laboratory stage of development. Examples of these are given in Table 8.5. However, many problems, such as the irreversible drift of the measuring signal, remain to be solved. Research into the development of amperometric sensors using microelectronic techniques is also of interest [267].

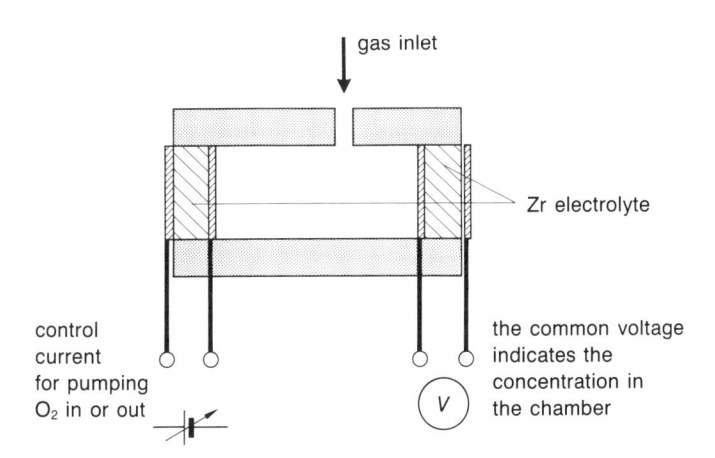

Figure 8.23 Coupling of potentiometric and amperometric O_2 sensors.

138 Chemical sensors

Table 8.5 Ion conductors used in solid electrolytes [252]

Ion conductor	Transported ion type	Components analyzed
stabilized zirconium dioxide	O^{2-}	O_2 (indirect H_2, H_2O_2, hydrocarbons)
doped thorium dioxide	O^{2-}	O_2
doped $SrCeO_3$	H^+	H_2, H_2O
polyantimonic acids $Sb_2O_5nH_2O$	H^+	H_2
$PbSnF_4$	F^-	F_2, O_2, H_2, NH_3
$K_2CO_3(+SrCO_3)$	K^+	CO_2
$K_2SO_4(+BaSO_4)$	K^+	SO_x

8.2.5 Chemically sensitive FETs (CHEMFETs)

In 1970 Bergveld [268] showed that it is possible to modify a well-known microelectronic component, the field-effect transistor (FET), for use as a chemical sensor, that is to say, to measure chemical concentrations in solutions. Instead of connecting a classic ion-sensitive electrode with a high-resistance FET amplifier, Bergveld's idea was to use the FET directly as an ion-sensitive electrode (ISE). This idea was to spawn many developments world-wide which are still continuing today. The preferred areas of application of this sensor are clinical diagnostics, process monitoring and control in the chemical, biotechnical and pharmaceutical industries and in environmental protection. Many variants of the principle proposed by Bergveld have since been developed. A classification proposed in [269] attempts to

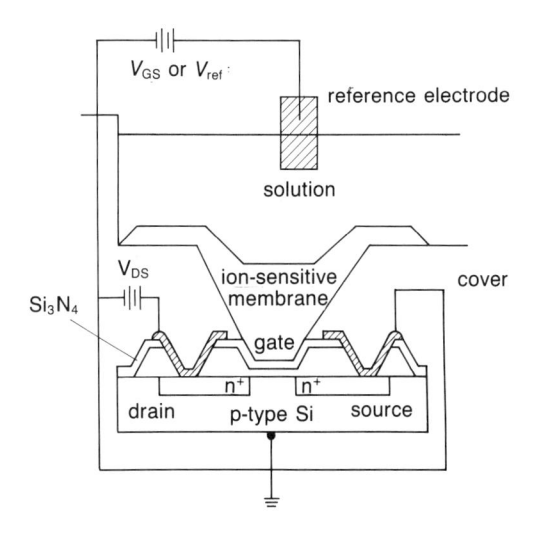

Figure 8.24 Structure of an ISFET.

impose some order on this diversity. In the following the most important basic principles are presented.

The longest-known and most frequently investigated component is the ion-sensitive FET (ISFET). This differs from a metal-isolator semiconductor FET (MISFET) in that the metal gate of the MISFET is replaced by an ion-sensitive material. Adsorption or reactions at this sensor layer modulate the gate potential of the transistor and thus the current flow between drain and source (Figure 8.24). The ISFET functions as an impedance transformer and amplifier directly at the detection site. The analytical information is provided by the height of the gate potential which controls the ISFET. This is composed of the voltage applied at the reference electrode V_{GS}, the jump in potential at the interface between the reference electrode and the analyte and the jump in potential at the phase boundary between the sensitive layer and the analyte. The chemical sensitivity of the ISFET is determined by the ion-sensitive material. In general, any membrane which is used in the ion-sensitive electrode can also be used as the gate material of an ISFET. The principle advantage of ISFETs over ISEs lies in the fact that the jump in potential at the ion-sensitive layer can be coupled capacitively and can be measured currentless. This means that it is possible to use ion-sensitive layers which are electrically non-conductive. In general, the following materials are used

- charge-impermeable surface-reactive membranes (e.g., insulating layers such as Si_3N_4, SiO_2, Ta_2O_5, ZrO_2);
- charge-impermeable inert membranes such as polymer layers (PTFE);
- charge-permeable, bulk-reactive membranes (e.g., ion exchange membranes such as PVC matrices with liquid ion exchangers);
- load-permeable, inert membranes (electrophoretic materials such as valinomycine);
- new membranes such as Longmuir–Blodgett (LB) films, implanted or modified SiO_2-Si_3N_4 layers, certain polymers (parylene).

The first pH sensors employed a standard FET with an SiO_2 gate insulator which performs the functions of an ion exchanger in the same way as the glass in a conventional pH or Na^+-sensitive electrode. However, the long-term stability of these components was inadequate. New production technologies led to considerable improvements in gate modification and encapsulation [271–273]. Extremely stable ISFETs for the measurement of pH values use Ta_2O_5 as the membrane [270, 274, 276] (Figure 8.25). Table 8.6 lists other methods of detecting alkaline, ammonium, sulphide and halide ions. Also of great importance and interest are new membranes which are based on special glasses and are used to detect heavy metal ions (Pb^{2+}, Cd^{2+}, Sn^{2+}, Tl^{2+}) in water [275]. The ISFET, which is generally in planar form, does not function stably as a single component. As in conventional ISEs, a reference electrode is still necessary for a complete measuring cascade and this electrode has to be miniaturized. The solution to this problem lies in the use of two ISFETs or an ISFET–MISFET combination (measuring or reference ISFET). This makes it possible to reduce interference effects (drift, influence of light, temperature). The

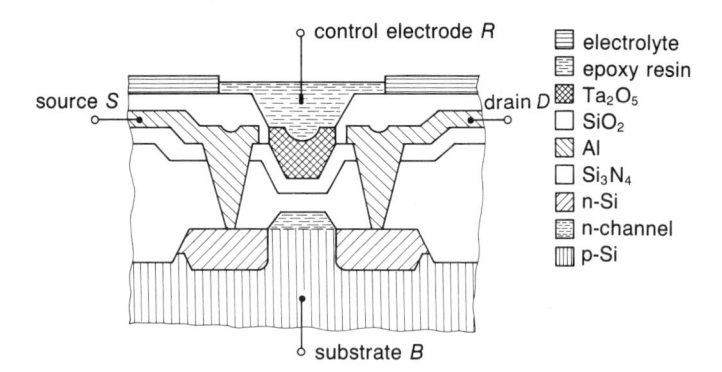

Figure 8.25 Structure of an n-channel ISFET with pH-sensitive Ta_2O_5 gate layer [270].

Table 8.6 Example membrane layers and the substances they can be used to detect

Membrane type	Sensitive material	Detectable substances
Dielectric	Si_3N_4, Al_2O_3, Ta_2O_5 Al-, B-, Na-Al silicates	H_3O^+ Na^+, K^+, Ca^{2+}
Crystalline	AgCl, AgBr, Ag_2S, LaF_3	AG^+, La^{3+}, Cl^-, F^-, Br^-, S^{2-}, ...
Heterogeneous	Ion exchangers polymer + ionophore, ion-implanted SiO_2, Si_3N_4	H_3O^+, Ka^+, Na^+, Cl^-, F^-, NO_3^+, NH_4^+

reference ISFET or MISFET is known as the REFET. This dual structure can be realized using CMOS or NMOS technology. The shared controlling electrode can then be used for any type of conductive compound. The conversion of the chemical signal into an electric signal is performed at the phase boundary between an ion-selective membrane and the electrolyte. The signal per decade of concentration change, that is to say, the chemical sensitivity, lies between 40 and 60 mV [272]. After suitable signal processing, for example using a differential connection as illustrated in Figure 8.26, measuring signals of a few hundred millivolts are obtained. The theoretical response times of ISFETs are of the order of a few microseconds. However, typical values lie between 10 ms and several hundred milliseconds.

Drift phenomena as well as the influence of temperature and light are undesired effects. Drift is particularly disruptive for long-term measurements which require high degrees of accuracy. It is primarily dependent on the membrane material and can be eliminated by continuous calibration or external correction. If the temperature is determined by on-chip measurement, then its influence can be corrected. Disruptive light influences can be avoided if the chip is appropriately constructed.

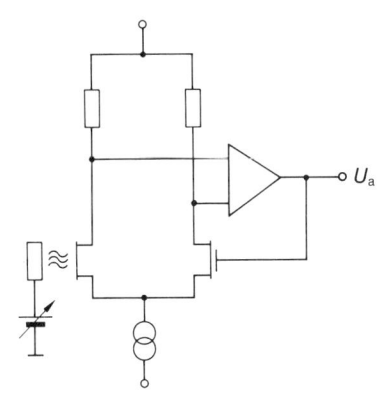

Figure 8.26 OPV with ISFET-MISFET differential input stage.

Alongside the ISFET, the above-mentioned suspended gate CHEMFET (see Figure 8.10) is also of interest in fluid systems since this device has an increased lifetime. In this sensor a polyimide grid is located approximately 1 µm above the gate area. This polyimide grid (10 µm perforations) can be considered as the support for the ion-selective layer. The grid improves the adhesion of the membrane to the FET and as a result increases its lifetime.

In another modification to the CHEMFET design, the gate is replaced by a sensitive enzyme layer. This type of device is known as an ENFET and will be dealt with in the discussion of biological sensors.

Despite considerable advances in the development of CHEMFETs, there are still a large number of problems to be solved, such as encapsulation, long-term stability, and drift. It has been shown that in principle a lifetime of approximately six months can be achieved [272]. This is of no interest for single-use sensors in medical applications. However, for repeated-use applications in the medical field (Figure 8.27), for analytical purposes or for industrial applications these sensors show much promise for the future.

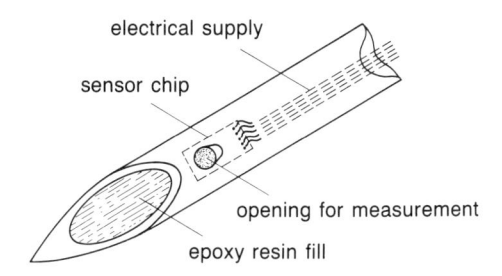

Figure 8.27 Example of a tubular sensor for insertion in tissue [272].

142 Chemical sensors

Integrated structures are increasingly attracting the attention of ISFET developers. These may consist of

- the integration of several ISFETs on a single chip: monofunctional and multifunctional [277, 278];
- the integration of electronic components for signal processing (amplifiers, differential amplifiers) or for the compensation of interferences (temperature sensors) [278, 279];
- the realization of microreference electrodes.

Figure 8.28 presents an example of a multisensor for the simultaneous detection of H^+, K^+ and Na^+ ions in a drop of blood [278].

8.2.6 Specific designs

8.2.6.1 Optrodes

Advances in fiber optic technology have made it possible to use optical fibers in chemical sensors. Such devices are known as *opto-chemical sensors* or *optrodes* [280, 281], and are based on an apparently very simple operating principle (Figure 8.29).

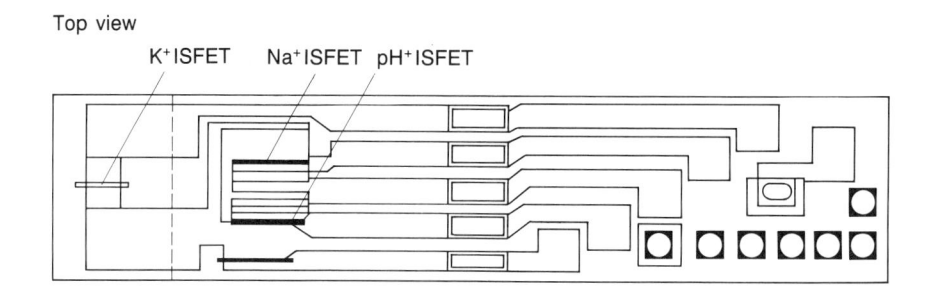

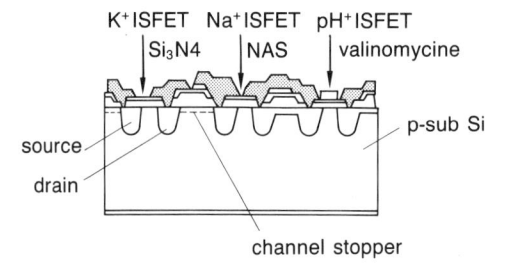

Figure 8.28 Structure of a multi-ISFET chip.

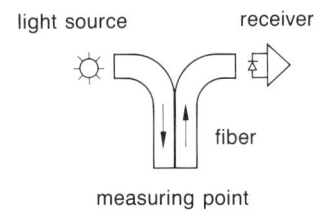

Figure 8.29 Chemical fiber optic sensor [281].

Light is passed along the optical fiber from a light source (semiconductor laser, light-emitting diode, halogen lamp) to the measuring point where it is absorbed, reflected or scattered. This can be performed either by the material under examination or as a result of a chemical reaction which is influenced by this material. The signal which has been modified in this way is passed via an optical fiber to the detector where it is focused, filtered and converted into an electric signal which corresponds to the measured value required for analysis. Apart from the general advantages of fiber optic sensors compared to other types, the merit of these optrodes resides in their selectivity and the absence of reference electrodes. I shall only present a few of the wide range of possible examples [282–287].

A simple example application for an extrinsic sensor is, for example, the measurement of blood oxygen saturation. The purple-red blood colourant haemoglobin is converted into the bright red oxyhaemoglobin through the addition of O_2. The change in reflectance can be used to determine the level of oxygen saturation in the blood.

A second generation of sensors makes it possible to measure analytical parameters such as pH values or O_2 and CO_2 partial pressures. However, these devices require a chemically selective sensing layer whose optical signal is supplied by stationary indicators which function as converters. Nowadays fluorescent materials are preferred as indicators because of their high degree of sensitivity and selectivity. Figure 8.30 depicts an O_2 optrode. In it the hollow end of the optical fiber is filled with a fluorescent pigment which is connected to small polystyrene spheres and is coated with an oxygen-permeable membrane. An optical fiber (3) transports blue excitation light to the particles at the fiber end. Green fluorescent light, whose intensity depends on the O_2 partial pressure, and blue scattered light (which is used

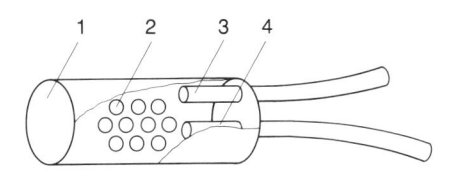

Figure 8.30 Fiber optic sensor for O_2 (1: permeable membrane; 2: polymer spheres with dye; 3: fiber for entry of light; 4: for exit of light) [232].

as a reference signal) pass along the second fiber (4) to the receiving diode. If the examined system contains O_2 then this causes the extinction of the fluorescence. The normed intensity of the fluorescence is inversely proportional to the O_2 partial pressure. This optrode can be used for clinical measurements, especially at low O_2 partial pressures.

In principle this type of continuous O_2 measurement can also be used for the measurement of oxygen levels in aircraft cabins for the rapid release of oxygen masks when pressure falls [280]. Another interesting principle makes possible the simultaneous measurement of blood pressure, pulse rate and O_2 and CO_2 pressure in the blood [288]. This is achieved by inserting a bundle of fibers in one end of a short tube (Figure 8.31). At the other end there is a flexible, gas-permeable, reflective membrane. Light of wavelengths 760 nm and 2 μm is emitted into the cell for the measurement of the O_2 and CO_2 partial pressure. Part of this light is specifically absorbed by the two gases and the remainder is reflected back along the fiber bundle. The distribution of light between the fibers depends on the deflection of the membrane. The frequency and amplitude of the light shift provide continuous information about the pulse rate and the blood pressure. Another example is illustrated in Figure 8.32. One end of each of the fibers in a catheter is provided with a layer of specific sensitivity. The fluorescent light provides information about the property to be measured. The built-in temperature sensor makes it possible to take account of the temperature dependence of all the sensors during the process of data analysis.

In principle, it is also possible to use the decay time of the fluorescence as a measure of the concentration of a species. The latest developments use only one optical fiber, at the measuring end of which a number of thin layers are superimposed. Multiplexed light of varying wavelengths is then passed along an

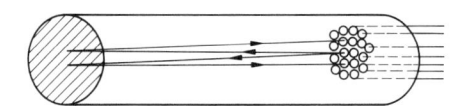

Figure 8.31 Combined sensor for the detection of gas contents and pressure [207].

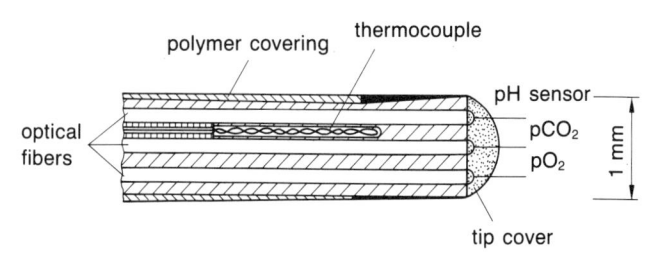

Figure 8.32 Miniature sensor for the continuous detection of pH, pO_2 and pCO_2 [280].

optical fiber and analyzed. This permits greater miniaturization. Optrodes also show great promise in the field of biological and immunosensors.

Table 8.7 groups together the typical types of optical gas sensor, while Table 8.8 lists sensors for the detection of specific ionic species.

8.2.6.2 Biosensors

The principles so far explained for chemical sensors can also be applied almost in their entirety to biosensors. Despite this, it is necessary to discuss them separately since, on the one hand, there is a wide variety of biological materials to be detected and, on the other, specific configurations have to be produced [289–294, 374].

A biosensor is a configuration in which a biologically sensitive element is connected to a physical converter and electronics (Figure 8.33). The objective is to generate a signal which is proportional in either its amplitude or its frequency to the concentration of the substance which is to be detected. The biologically sensitive element is a bioactive sensor and is known as the *receptor*. The substance to be detected, the analyte, binds to this receptor. This presupposes that the analyte interacts with the receptor, where it is chemically converted, or is influenced in some other way. Thus the first step in the design of a biosensor is to ensure that the

Table 8.7 Selected gas optrodes [280]

Species	Detection method	Typical measuring range
Oxygen	Fluorescence extinction	0.5–300 torr
	Phosphorescence extinction	0.0005–0.1 torr
CO_2	via pH measurement in internal buffer	0.5–100 torr
SO_2	Fluorescence extinction	70–50 000 ppm
Chlorine	Internal absorption	0.01–0.05
Hydrogen	Interferometric measurement	5–2000 ppm

Table 8.8 Selected sensors for electrolytes

Species	Detection method	Typical measuring range
H^+	Reflectometry	2.5–3 pH units
	Fluorescence	4–5 pH units
Cu^{2+}	Internal absorption at 820 nm	50–500 mM
Al^{3+}	Fluorescence of morine complexes	1–100 µM
K^+, Ca^{2+}	Optical potential measurement	0.001–100 mM
	Reflectometry	5–50 mM
Na^+	Fluorescence	20–200 mM
Halides	Fluorescence extinction	15–200 mM

146 Chemical sensors

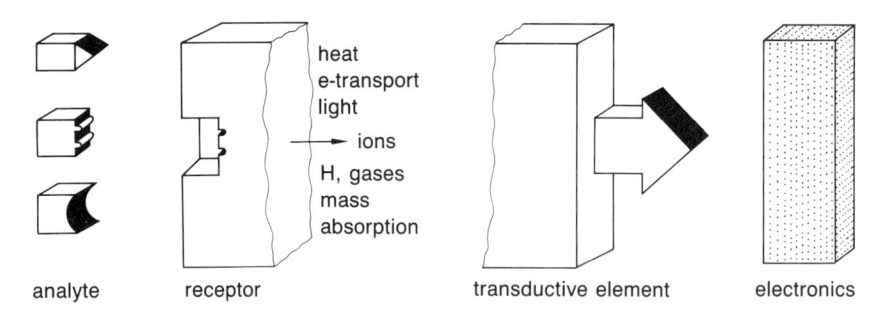

Figure 8.33 Structure of a biosensor.

receptor binds specifically with the analyte. This can cause the charge distribution in the receptor component to change. However, unwanted secondary reactions may also occur. This problem can be avoided by displaying the products of any consequent chemical reactions using a catalytic sensor. It is necessary to connect a suitable converter for the generated effect, that is, variation in charge, gas or ion generation, or heat change. This is the second step in the development of a biosensor. The converter can take the form of an ISE [298], an amperometric electrode [295], an FET [294], a conductivity sensor [299], a thermistor [300], an FOS [280] or a piezoelectric crystal [301]. Signal preparation then takes place in a third stage and may employ procedures to compensate for drift and temperature dependence or to perform continuous calibration. This considerably extends the sphere of application of biosensors, which otherwise frequently exhibit a high degree of thermal and chemical instability.

Having presented this general definition of biosensors, I shall now give a few selected examples. As a branch of sensor technology, biosensors are currently undergoing considerable development, and many interesting areas of application either exist already or have been identified for the future (medicine, biotechnology, process control, agriculture, chemistry, environmental protection, pharmaceuticals, etc.). Most current development work is concentrating on questions concerning the receptor material and ways of fixing it to the converter. The former requires considerable biochemical knowledge, the latter raises complex technological questions [290]. In contrast, known sensor principles can be applied to the design of the converter.

The first biosensor was the *enzyme electrode* described in [295]. Clark designed this sensor by placing an enzyme solution with a semi-permeable membrane in front of the O_2 electrode which he had also invented (see Section 8.2.3). This enzyme electrode technique continues to dominate biosensor technology today. If, for example, blood sugar levels have to be detected then the enzyme glucose oxidase (GOD) is used. This enzyme is a catalyst for the oxidation of glucose to gluconic acids and hydrogen peroxide:

$$\text{D-glucose} + O_2 \xrightarrow{\text{GOD}} \text{gluconolacton} + H_2O_2 \tag{8.11}$$

The resulting H_2O_2 can be observed at the electrode through a process of anodic oxidation. Glucose concentrations of between 10 µmol/l and 5 mmol/l can be measured. The electrode described by Clark is an amperometric enzyme electrode. Potentiometric enzyme electrodes can be used to measure analytes whose enzymatic transformation causes a change in the pH value, the creation or consumption of ions or the production of gases. Such electrodes are used, for example, to determine the presence of urea or penicillin. The literature so far contains descriptions of biosensors for determining approximately 110 different substances, among them substrates, inhibitors, activators, cofactors, prosthetic groups, enzyme activities, haptenes, antigens or micro-organisms [291]. Table 8.9 lists some of these [296]. Table 8.10 lists the principles on which biosensors can be based. Alongside enzyme

Table 8.9 Examples of bioelectrodes for specific analytes [331]

Analyte	Receptor	Sensor
Glucose	Glucose oxidase	pH, O_2, H_2O_2 electrode
Ethanol	Alcohol oxidase	O_2 electrode
D-amino acids	D-amino oxidase	NH_4^+ electrode
L-glutamine	L-glutiminase	NH_4^+ electrode
Penicillin G	Penicillinase	pH electrode
Urea	Urease	NH_4^+, pH electrode
Hepatitis B	Anti-H BsAg Antihepatitis Surface antigen	GOD + O_2 electrode Ag/AgCl electrode
Peroxidase	Con A	H_2O_2 electrode

Table 8.10 Principles of biosensors

1. Affinity sensors $S + R \leftrightarrows SR$		2. Metabolic sensors $S + R \leftrightarrows SR \rightarrow P + R$	
Change in electron density		Substrate usage and formation of products	
Receptor R	Chemical signal S	R	S
Dye	Protein	Enzyme	Substrate
Lectin	Glycoprotein	Organelle	Cofactor
Apoenzyme	Prosthetic group	Cell	Effector
Antibody	Antigen, haptene	Tissue tear	Enzyme activity
Receptor	Hormone		
3. Coupled and hybrid systems		4. Biomimetic sensors	
Sequence		Receptor R	Physical signal S
Concurrence			
Anti-interference		Carrier enzyme	Sound
Amplification			Strain
			Light

electrodes, micro-organisms, antibodies, lectins and biotic receptor systems are used for the detection of molecules in biosensors.

The next problem is to fix or immobilize these materials at the electrode. In principle, this can be achieved by enclosure methods, three-dimensional linking, covalent bonding or adsorption. Figure 8.34 presents a number of variants [296]. The layer containing the embedded biological component must fulfil certain conditions such as:

- adequate adhesion to the substrate;
- suitable degree of semi-permeability;

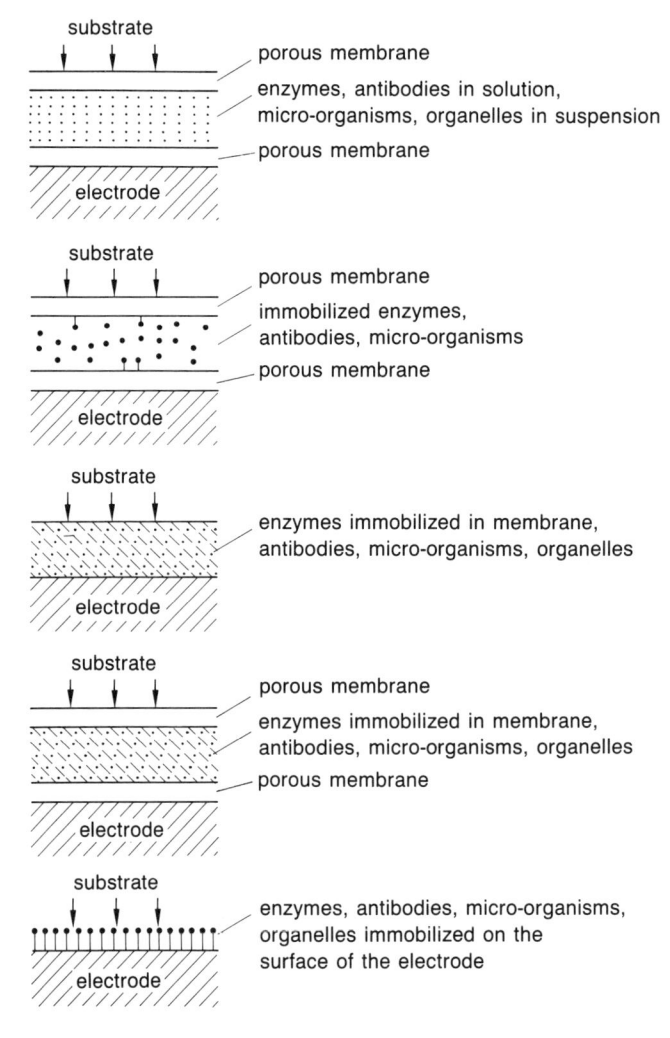

Figure 8.34 Possible configurations for bioelectrodes [296].

- reproducibility of the chemical reaction;
- operating period of at least a few days (with the exception of disposable sensors);
- storage stability of several months;
- structuring must be possible.

Biospecific electrodes are classified into three generations of development depending on the degree of integration of their components. The simplest principle is based on the coupling of biocatalysts which are either included in membranes located in front of the electrode, which functions as a transductor, or are bound to the membrane (first generation). Direct physical or chemical bonding to the surface of the electrode results in rapid response times due to the omission of the dialysis membrane. Coimmobilization of cosubstrates makes it possible to extend the dynamic range of enzyme electrodes (second generation). Fixing the biocatalysts to an electronic component, for example the gate of a FET, which performs signal processing directly, leads to further miniaturization and makes it possible to house a number of sensors on a single chip (third generation). It is thus possible to manufacture sensor systems which can detect various analytes simultaneously or which can identify the concentration profile of a single analyte. Although there is no doubt that this generation will be of great importance in the future, problems still exist today. These include the difficulty of applying a technologically compatible, structurable layer to a pH ISFET or the problem of minimizing side-effects such as the influence of changes in pH values or the buffer capacity of the measuring probe. For this reason first-generation sensors continue to predominate today, with amperometric enzyme electrodes based on multilayer membranes enjoying the widest range of application. These possess a linear measuring range over several orders of magnitude of concentration with a lower detection limit of 1 µmol/l, a reproducibility of 10 000 measurements per membrane, and permit a probe frequency of 300/h [297]. Applications in clinical diagnostics, for example the Enzyme-Chemical Analysor ECA 20 for the detection of blood sugars, in fermentation control or in environment protection are impressive examples of the wide range of possibilities.

The main advantage of biosensors lies in the fact that they can be used directly in liquid environments and possess a high degree of selectivity which makes it possible to measure individual components in a complicated mixture without having to isolate them first. Recent work has investigated the measurement of analytes in the gas phase, biocides for example. This is achieved by immobilizing enzymes or antibodies on piezoelectric crystals [301]. If analyte bonding causes the weight of the crystal to rise, this results in a change in the oscillating frequency. The first attempts to use this method to detect drugs and explosives were successful. This technique will be discussed more fully in connection with resonance sensors. SAW sensors, which are becoming increasingly interesting as both chemical and biological sensors, will also be dealt with under this heading.

150 Chemical sensors

8.2.6.3 Humidity sensors

The ability of air to hold water has a considerable influence on a large number of processes which proceed in the normal atmosphere. In terms of the number of applications involved, water is possibly the most important substance in our daily life and it occurs in air, solids and fluids. It has to be detected in these materials. Although the term 'humidity' is usually understood to refer to the water content of the air (which represents the most important task of humidity measurement), it is often useful to be able to determine the water content of solids or liquids directly.

When specifying the concentration of water vapour in gases, principally air, it is important to distinguish between:

- *absolute humidity*, which specifies the amount of water vapour present per unit volume of the gas, and is measured in grams per cubic metre;
- *saturation humidity*, which specifies the maximum amount of water per unit volume of the gas that the gas can hold at a given temperature; and
- *relative humidity*, which is the ratio of absolute humidity to saturation humidity and has a value between 0 and 1.

Of use also is the ratio of the partial pressure of the water vapour at the measuring temperature to the possible saturation pressure at the same temperature. In general, it is relative humidity that represents the most important measured value. A frequently used measure is the *point of condensation*. This is the temperature at which the absolute humidity of the atmosphere which is being considered is such that the relative humidity assumes the value 1. When the temperature sinks below this point the water vapour starts to condense.

The direct measurement of the water content of liquids and solids is more difficult since it is rarely possible to specify the water content of a product as an isolated measurement. In solids this value is easy to determine by weighing the product, drying it and then weighing it again. However, there are a number of sources of error connected with this procedure, for example disintegration of the probe, the length of the drying period, and the type of water bonding.

Reliable measuring systems have long existed for the determination of humidity levels [245]. These include mechanical procedures such as the hair hygrometer, the psychrometer, and the LiCl humidity detector in which the surface resistance is measured. The structure of this classic detector is illustrated in Figure 8.35. An a.c. voltage is applied at the electrodes numbered 3. This causes a current

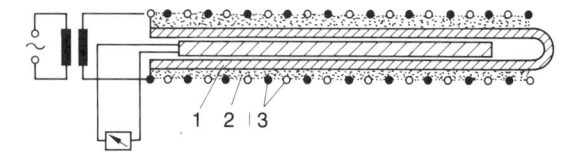

Figure 8.35   LiCl humidity gauge (1: measuring resistor made of Pt; 2: glass fabric with LiCl; 3: electrode filaments).

to flow through and heat the LiCl solution. Water consequently evaporates from the solution. As soon as all the water has evaporated, the conductivity, and with it the current between the electrodes, falls off sharply and the temperature drops. The hygroscopic LiCl is now able to accept water from the air. Its conductivity increases and the current again causes water evaporation. In this way the temperature regulates itself to a state of equilibrium between the electric power supply and the thermal energy required for evaporation. This equilibrium is dependent solely on the water vapour pressure of the surrounding air and is therefore a measure of the absolute humidity. The temperature at equilibrium is recorded by a measuring resistor (1) and is subsequently processed as an electrical quantity. It is possible to measure relative humidities of 15–90% at temperatures of 0–60°C. The response times are of the order of minutes. The technical significance of these classical detectors is today giving way to miniaturizable sensors which are cheaper, faster and sometimes more accurate. There are three avenues of development.

Changes in resistance, particularly in surface resistances, form the basis of one type of sensor. This includes the ceramic hygrometers which also make use of the adsorption of water at the internal surfaces of porous ceramic materials which have been sintered from powder. The ceramics used are TiO_2-V_2O_5, $MgCr_2O_4$-TiO_2 [227], $ZnCr_2O$-$LiZnVO_4$ and perovskite [303]. Sensors made from $MgCr_2O_4$-TiO_4 are used commercially in microwave ovens [304]. They have a response time of approximately 20 s and measure humidities in the range 30–90%. Other resistance-based sensors consist of sulphonated polystyrene or carbon powder suspended in gelatin cellulose. The surface conductivity of these sensors changes when they accept water. Materials such as LiF/Al_2O_3 composites [305], zirconium phosphates and silicates [306], polysiloxanes with hydrophilic groups [308] and certain polymers hold a certain promise for this class of sensor. Polymers should be moisture-sensitive and at the same time insoluble in water. Linked polyvinyl pyridine currently meets these requirements best.

A second type of sensor makes use of changes in capacitance. In general, these sensors record a wide range of humidities and are more accurate than the sensors based on the resistance method. For the sake of simplicity, if we ignore marginal effects the capacitance of a plate capacitor is given by

$$C = \varepsilon_r \varepsilon_0 A/d \tag{8.12}$$

where A is the area of and d is the distance between the plates, ε_0 is the dielectric constant (DC) and ε_r is the relative DC. ε_r determines the sensor effect. This increases as the DC of the carrier material is reduced. Initially porous Al_2O_3 with a relative DC of 10 was used for the sensor material. Thin-film methods are used to apply it to the substrate (glass, ceramic) since they permit simple structuring and integration of the layers. Alongside Al_2O_3, tantalum oxide and titanium oxide are used. Recently polymers with a relative DC of 2–15 have been increasingly selected. These materials possess a high degree of long-term stability and include cellulose acetate [309], polystyrene, polyimides which can be formed into sensitive layers

152 Chemical sensors

using cast coating methods, as well as polymers produced by glow discharge polymerization [310]. Capacitance can be measured either in the plane of the film-like adsorber (Figure 8.36) or perpendicular to this (Figure 8.37). In the second case, the 'sandwich' configuration, one of the two electrodes must be permeable to moisture. In general, this is achieved by using a gold film whose thickness represents a compromise between the stability of the element (thick-film) and low response times (thin-film).

This type of polymer-based humidity sensor is available, for example, for the measurement of relative humidities across the entire range of values within temperatures between $-80°C$ and $175°C$ and possesses an accuracy better than 1% [311]. When operating near the condensation point, these sensors can be used at temperatures between $-60°C$ and $+30°C$. In this example the evaluation electronics consists of an HF measuring bridge with subsequent linearization of the signal values. Other applications make use of ASICs [309].

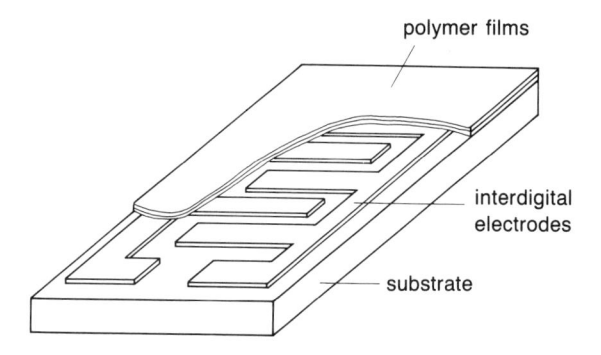

Figure 8.36 Planar capacitive humidity sensor.

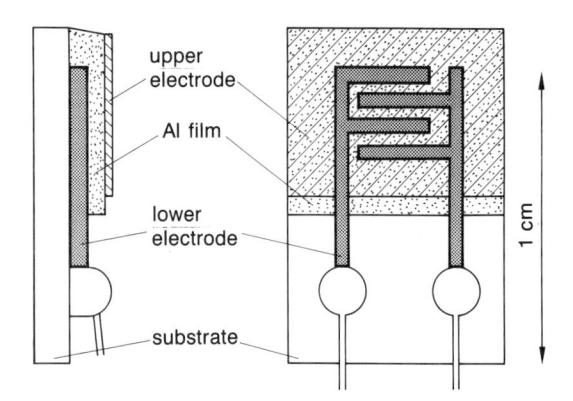

Figure 8.37 Humidity sensor using 'sandwich' configuration.

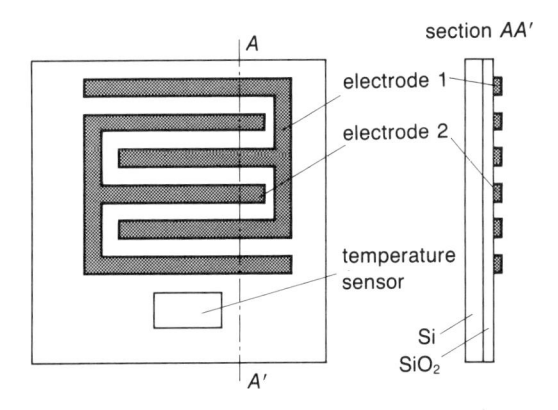

Figure 8.38 Condensation point sensor.

Capacitive, polymer humidity sensors can now be manufactured with low drift and a lifetime of several years. A high degree of reversibility can be obtained. They can sometimes be used to determine the water content of liquid media such as organic solvents or fuel [232].

The third type of humidity sensor is the condensation point sensor. The condensation point can be measured, as its definition suggests, by cooling a test surface and observing condensation or the formation of a liquid layer as a function of temperature. This 'observation' can, for example, be performed optically. If the test surface is smooth and reflective, it is dulled by the deposition of water and the reflection of a light beam becomes diffuse. This effect is easy to detect. More frequently, capacitive or conductive measuring methods are used. Figure 8.38 shows the example of a capacitive condensation point sensor. An electrode applied using thin-film techniques functions as a capacitor with a capacitance which changes when liquids settle on it.

There are no miniaturized sensors for measuring the water content of solids. Classical methods of measuring electric conductivity predominate, together with microwave and infrared absorption.

9 Sensors based on 'classical' measuring elements

The principles behind many measuring elements have long been known and it is difficult suddenly to describe these as 'sensors'. Such measuring devices include temperature gauges based on the resistance thermometer, the thermocouple, or on PTC or NTC elements or, again, the WSG with its many different geometrical design types, manufacturing techniques (film, wire, semiconductor WSGs) and innumerable applications. There are many detailed accounts of such devices, and these will undoubtedly continue to play an important role in the future [3, 245, 312, 313]. It should also be recognized that microelectronic technology has made possible the development of new techniques (the monolithic or thick-film temperature sensor, the capacitive pressure sensor and the integrated WSG). There is no doubt that micromechanics, too, has opened up new possibilities. Moreover, information processing microelectronics is playing a not inconsiderable role as a stimulus for the development of 'classical' measuring elements. If the output of this type of sensor provides an electric signal as a piece of integral data, then appropriate algorithms and corresponding software can retrieve information which used to be inaccessible. This category of sensors includes inductive, capacitive, ultrasound, eddy current, correlation and force-moment sensors. Most of these cannot be miniaturized and are therefore not 'microelectronic' sensors. However, they do satisfy our original definition of the term 'sensor'. Some current examples of this type of sensor will now be presented.

9.1 Inductive sensors

Changes in induction have long been employed for measurement purposes. In its simplest configuration (a combination of coil, magnet/inductance core and object), the inductive sensor is based on the fundamental principles of the law of induction:

$$U_{ind} = -N \, d\Phi/dt \tag{9.1}$$

where U_{ind} is the induced voltage in the coil, $d\Phi/dt$ is the change in magnetic flux over time and N is the number of times the coil is wound around the core. The change in flux can be caused in a wide variety of ways, through rotation or relative movement of a magnet or through displacement of the core. If a second magnet is introduced into the configuration then the inductivity is affected (self-induction/mutual induction). In general, we can write for the induction L:

$$L = L(N, A, \mu) \tag{9.2}$$

where A describes the dimensions of core and coil and μ is the permeability of the core. A practical sensor can be realized by the skilful selection of a parameter while maintaining the other parameters constant. There are a number of possible variants of sensor. The most frequently encountered is the *inductive displacement transducer*, the principle behind which is illustrated in Figure 9.1. A usually rotationally symmetric sensor housing accommodates a primary coil which is supplied with an alternating current of a few kilohertz by an oscillator. At the same time, the housing contains two secondary coils connected in series. An alternating voltage, which depends on the position of the core, is induced in the 'direction sensing' core in the coil system. After demodulation and amplification, the analog output is the differential signal of the two secondary coils. Such a device is known as a *linearly variable differential transformer* (LVDT) sensor. This type of inductive sensor or displacement transducer is manufactured in many countries. Today it is still used

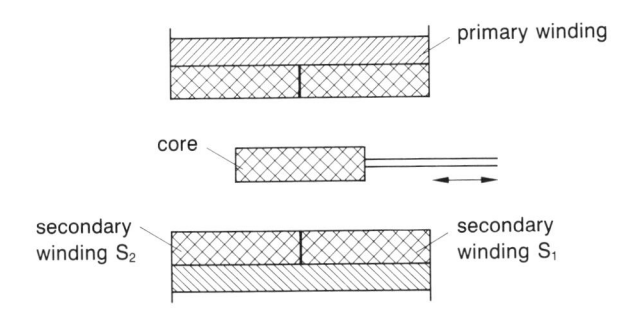

Figure 9.1 Inductive displacement transducer.

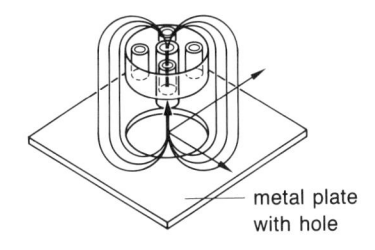

Figure 9.2 Inductive sensor for multidimensional measuring tasks.

to measure displacements ranging from fractions of a millimetre to several metres. They can be manufactured to meet stringent accuracy and resolution requirements (for example, accuracy better than 0.2%, high linearity, operating temperature range from −50°C to several hundred degrees, frequency range to 15 kHz). A modification to this principle produces the inductive sensor. Inductive sensors are

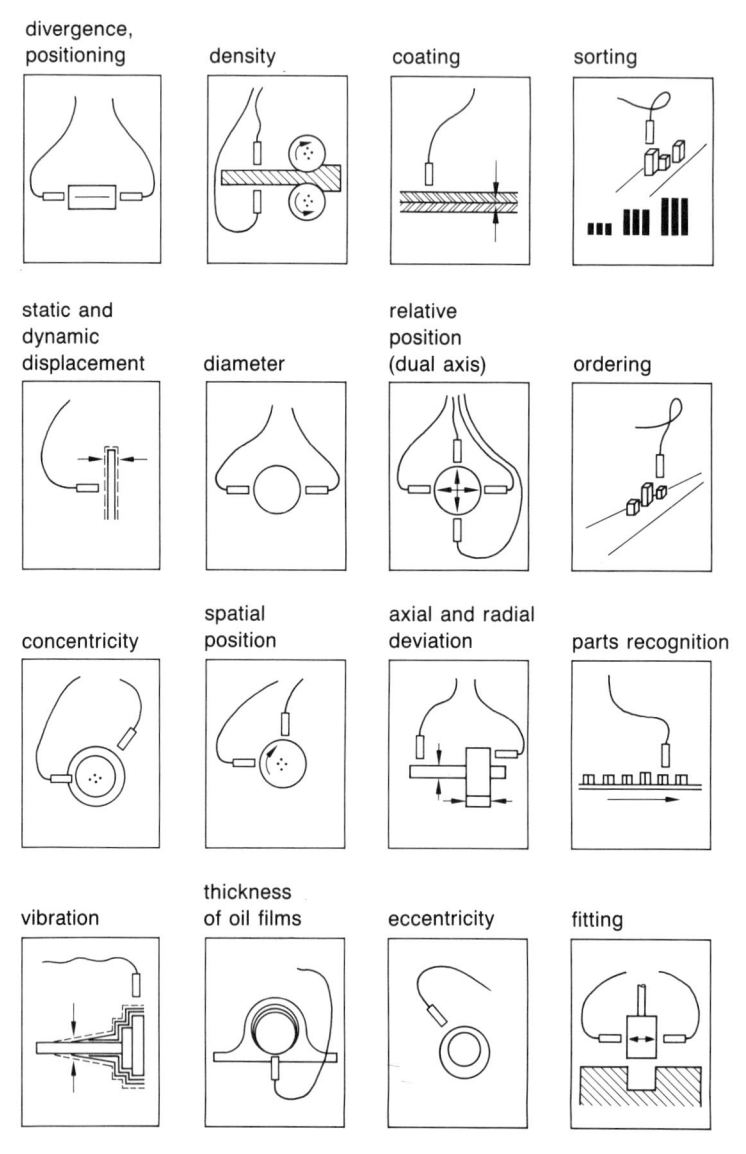

Figure 9.3 The use of inductive sensors for various measuring tasks.

also used for measuring speeds of rotation. Such a device may consist of a gear wheel made of ferromagnetic material which moves past a coil with a ferromagnetic core. The advantage of this type of sensor is that it needs no auxiliary power source. The drawback is that the amplitude of the signal is frequency-dependent and that the device is highly sensitive to vibrations in the individual components. Inductive sensors are important in the field of robotics. They are used for the recognition of joins, edges or distances [314]. Figure 9.2 presents a sensor which is also of interest in robotics. It consists of a transmitter coil and four receiver coils. The stray field generated by the transmitter coil induces an equal voltage in the four receiver coils which are arranged symmetrically around it. This voltage is uniformly modified by a metal surface which runs parallel to the measuring device. The sum of the receiver coil voltages can be used to calculate the distance between the sensor head and the metal surface. This type of sensor can detect deviations from the centre of the hole (Figure 9.2) to an accuracy of approximately 1.2 mm. One example application is the fitting of rubber plugs to vehicle bodywork.

Inductive sensors are also used as pressure sensors in which the coil core can be located directly on the membrane. They are also used as torque gauges [315]. Figure 9.3 presents an overview of the many possibilities.

9.2 Capacitive sensors

The operating principle underlying these sensors employs the dependence of the capacitance on either the dielectric constant, the plate area A or distance d. Figure 9.4 shows a number of different configurations. The main advantages of capacitive sensors lie in their simple construction, the almost perfect degree of calculability they offer and their very high resolution. They can be used at high temperatures and yield a frequency-analog output signal. Their disadvantages include the small measuring signal (sensitivity to interference) and the influence of cable capacitances. Capacitive sensors have long been used as distance, position, fill level and pressure sensors [316, 317]. Thanks to the use of thin- and thick-film methods it is now also possible to manufacture miniaturized sensors. Sensors produced using micromechanical techniques are of special interest as acceleration sensors.

An apparently very simple but very important type of capacitive sensor is the *dielectric field sensor* [318]. This sensor consists of three electrodes (Figure 9.5). Antiphase alternating voltages are applied at electrodes I and II. The resulting potential is recorded at the site of the third electrode S, and is then subjected to frequency- and phase-selective further processing. In the balanced, undisturbed state, the electrode S is at zero potential. If an object (conductor or non-conductor) penetrates the measuring bulk the ratios of the potentials change. A direct or a direction-dependent distance measurement can then be performed depending on the position of the electrodes. The system also reacts with great sensitivity to changes in the sensor geometry. Over small ranges it is possible to assume an approximately

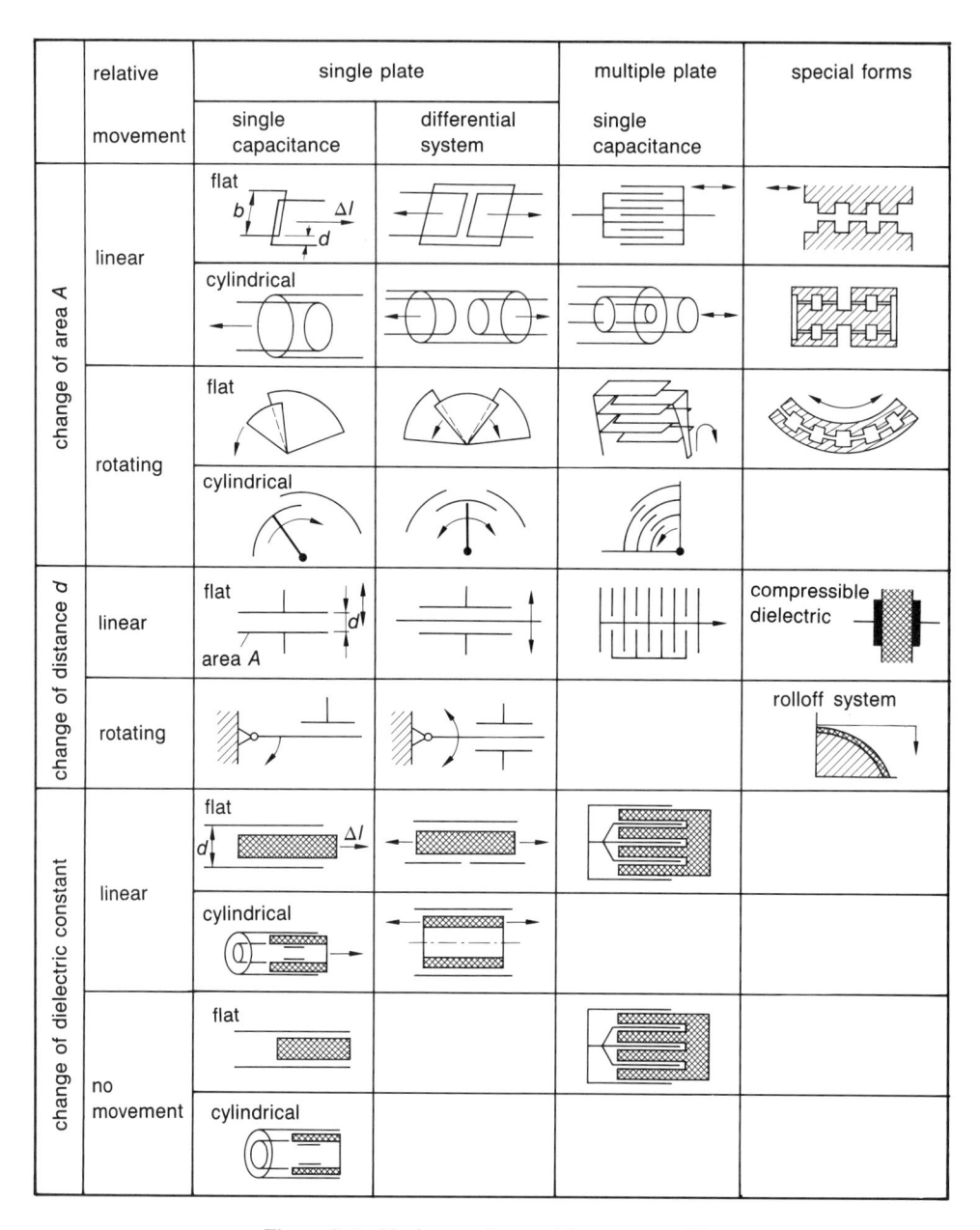

158 Sensors based on 'classical' measuring elements

	relative movement	single plate – single capacitance	single plate – differential system	multiple plate – single capacitance	special forms
change of area A	linear	flat; cylindrical			
	rotating	flat; cylindrical			
change of distance d	linear	flat (area A)			compressible dielectric
	rotating				rolloff system
change of dielectric constant	linear	flat; cylindrical			
	no movement	flat; cylindrical			

Figure 9.4 Variants of capacitive sensors [3].

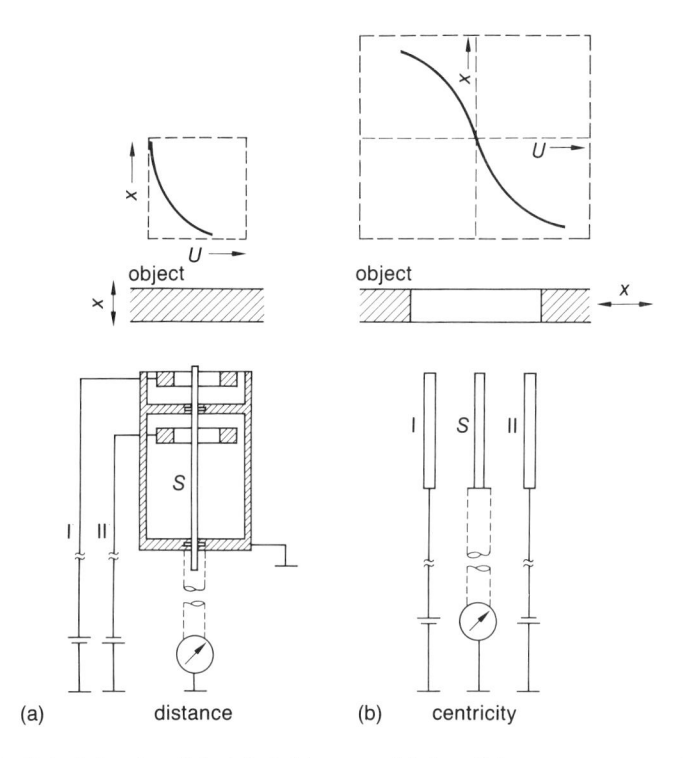

Figure 9.5 Principle of a dielectric-field sensor (a) for distance measurement; (b) for determining centricity.

linear distance dependence. In both cases the sensitivity of the system is 10^{-5}. Given the appropriate ceramic carrier materials, this measuring system can be used at temperatures of up to 1000°C. Measurable distance ranges from metres (at resolutions of centimetres) to millimetres (at resolutions of hundreds of nanometres) can be achieved. As a result, these sensors have a wide range of applications such as machine-tool supervision in metalworks or as presence and position sensors in robot arms, as well as in high-temperature applications.

9.3 Ultrasound sensors

The term 'ultrasound' is used to refer to the propagation of elastic waves at frequencies above 20 kHz. Nature has provided a wide variety of ways of generating and receiving ultrasonic waves. Technological development has allowed humans to add new methods. Historically, it is the piezoelectric effect that has played the decisive role, and this is still the case today [319]. When certain crystals are

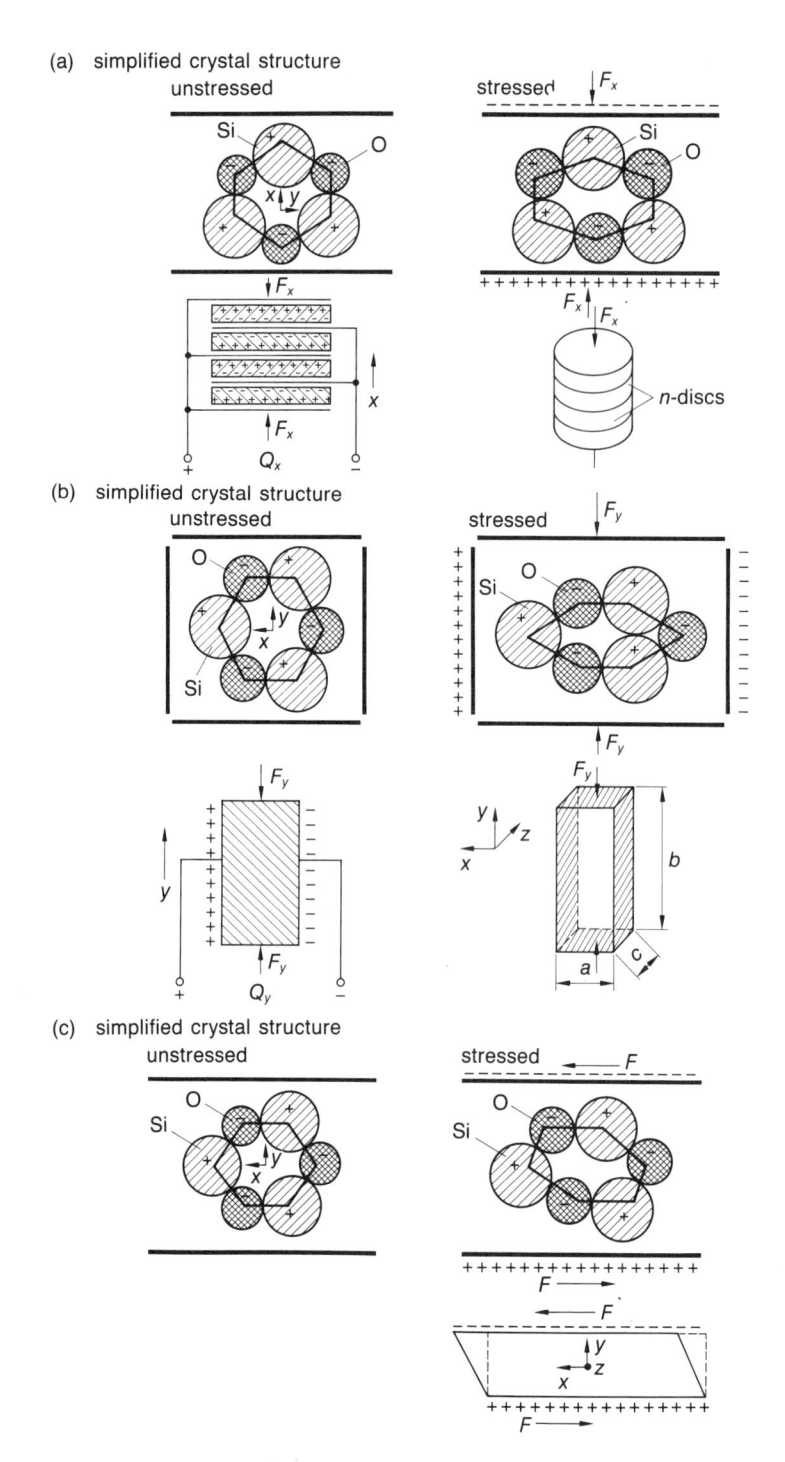

Figure 9.6 Various forms of the piezoelectric effect: (a) longitudinal; (b) transversal; (c) shear effect [3].

subjected to mechanical stress, electrical charges are generated. This phenomenon is known as the *direct piezoelectric effect*. These crystals are characterized by the presence of a polar axis or the absence of a centre of symmetry. Quartz is the best known representative of this class (Figure 9.6) [320]. The direct piezoelectric effect is reversible and is therefore used in sound generation. If an alternating electric field is applied to the quartz then mechanical oscillations can be generated and these are propagated as sound waves in an adjoining medium. Alongside quartz, piezoelectricity is a property of piezoceramics such as barium titanate or lead zirconium titanate, or substances such as lithium sulphate or lithium niobate. It should be possible to use transduction elements made of these materials in liquids, gases and solids. However, there is a considerable difference in the behaviour of these materials. If this type of transduction element is stimulated by a pulse it oscillates at a resonant frequency which is dependent on the speed of the sound and the density of the body. Only part of the sound energy crosses the transducer–medium interface and enters the bounding medium. This can be described in terms of the transmission factor T:

$$T = 2Z_M/(Z_M + Z_p) \tag{9.3}$$

where Z_M and Z_p are the sound impedances of the medium, $Z_M = \rho_M c_M$, and of the piezoelectric material, $Z_p = \rho_p c_p$. At the piezoelectric transducer–water interface, T is approximately equal to 0.8 and at the interface with air it has an approximate value of 10^{-5}. It is clear that, in comparison with the water interface, only a small proportion of the energy is transmitted from the piezoelectric body to the surrounding air. It is possible to improve the efficiency of energy transfer by applying impedance adaptation layers to the transmission side of the piezoelectric body. Piezoelectric transducers possess a high quality factor. When such a device is stimulated it exhibits a long decay time. It is only possible to generate very short sound pulses if the oscillator is damped. This in turn reduces the amplitude of the oscillations. If an undamped transducer is stimulated at the resonant frequency, the high quality factor results in corresponding, high-intensity ultrasound radiation. This advantage of piezoceramic oscillators is also found in the sensitivity of the otherwise narrow-band receiver-driven transduction element.

A new, very interesting material for ultrasound sensors is PVDF (see Chapter 10) [321]. This is a stretched, polarized polymer film with a thickness of a few tens of micrometres. This can be used to produce a variety of transducer geometries with a wide bandwidth. The acoustic impedance is smaller than that of piezoceramic materials ($Z_{PVDF} = 4 \times 10^6$ kg s^{-1} m^{-2}, $Z_{piezoceramic} = 2 \times 10^7$ kg s^{-1} m^{-2}) and is well adapted to use in biological media. Although the electromechanical coupling factor which describes the conversion of sound energy into electrical energy is smaller than in piezoceramics, this polymer film is being increasingly used as a receiver in both liquid and gaseous media. The main reason for this is the advantages conferred by wide bandwidths. It has also recently been used as a material for transmitting transducers, for example as a transduction element for airborne sound [322].

Another important category of ultrasound sensors comprises the *electrostatic transducers*, also known as *Sell transducers*. In principle, these are constructed in the same way as plate capacitors (Figure 9.7). One electrode consists of a massive, rigid plate, while the other is formed by a conductive metal layer which has been applied to a thin membrane. There is an air cushion between the film and the electrode. During operation a constant bias voltage is applied to the capacitor. This causes the film to be drawn to the plate with a defined force. A superimposed alternating voltage with an amplitude smaller than that of the direct voltage causes this force to change, with the result that the position of the membrane also changes. The air between the electrode and the film acts as a spring against which the mass of the membrane works. The spring–mass system thus formed has a resonant frequency which is dependent on the membrane and the air cushion. Through mechanical manipulation of the fixed electrode it is possible to adjust the resonant frequency of the system. The smoother the surface is, the greater the bandwidth of the transducer and the shorter the response time.

Although ultrasound techniques have been long established in many sectors, such as non-destructive materials testing, and are being increasingly employed in medical applications, an example being ultrasonic tomography, it is only in recent years that ultrasound sensors have started to play an important role. A reason for this has been the lack of suitable ultrasound transduction elements which can survive in demanding industrial environments and which operate at frequencies above the level of industrial noise. The influence of information processing microelectronics has again been an important factor. Thus the use of suitable algorithms has made it possible to extract new process information from data provided by ultrasound sensors. A few example applications should illustrate their significance. These include the measurement of distances, the industrial detection and classification of objects, and process measurement.

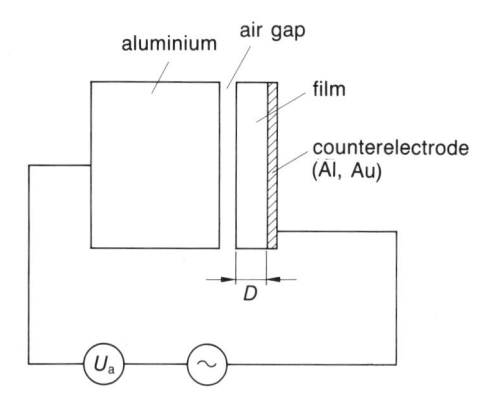

Figure 9.7 Electrostatic transducer [326].

Ultrasound sensors 163

Two parameters are central to the propagation of an ultrasonic wave. In the following these are illustrated using the example of alternating sound pressure $p = p(x, t)$:

$$p = p \exp(j\omega(t - x/c))\exp(-\alpha x) \quad (9.4)$$

namely the speed of the sound c and the sound absorption α. x is the distance travelled by the sound, t is the time and ω is the angular frequency. The properties of the medium through which the sound passes influence both c and α.

Ultrasound sensors as separation or distance sensors are primarily used in air in pulse–echo mode. A sound pulse is emitted by a transmitter, is reflected by the object to be measured and is received by the same or another transducer. The travel time t is used to determine the distance

$$l = ct \quad (9.5)$$

The same principle can be used for measuring fill levels, in which case the ultrasound may pass not just through the air but also through the solid or liquid. Many variants of this type of distance sensor are commercially available [323, 324]. One of the best-known examples is the Polaroid ultrasound distance measuring system for automatic cameras [325] which makes use of electrostatic sensors. Figure 9.8 illustrates the transmission–reception principle. The minimum distance that can be measured is 11 mm. In normal operation, objects at a distance of up to 10.7 m

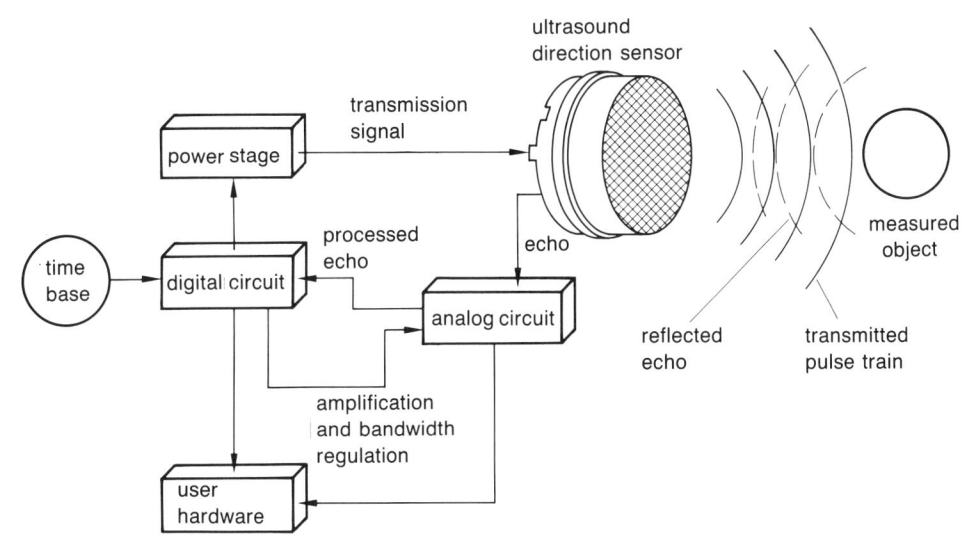

Figure 9.8 Transmit–receive operation of a Polaroid ultrasound distance measurement system [325].

164 Sensors based on 'classical' measuring elements

are detected. In general, the measuring range, resolution and accuracy of this type of distance sensor depend on the frequency and geometry of the sound transducer. High-frequency transducers focus the sound bundle better and thus achieve greater accuracy. However, the attenuation is greater and the distances that can be measured are shorter. In general, this type of transducer is used in the frequency range 20–400 kHz. When they are used it is essential that proper account is taken of environmental parameters which can influence the measurement, such as temperature, pressure, humidity or air pollution.

Ultrasound sensors are also used as presence sensors. A yes–no response concerning the presence of an object is made if a sound path exists between transmitter and receiver and a given body is or is not present in this path. More complicated is the use of ultrasound sensors to detect and classify objects, although in the future this will certainly be an extremely interesting area of application [326]. In contrast to optical sensors, which record surface profiles with great accuracy, ultrasound sensors record three-dimensional profiles. Excellent results can be obtained by coupling these two sensor principles, that is to say, by coupling ultrasound matrix arrays with a video camera. The high sensitivity of piezoelectric transduction elements makes them particularly well suited for the construction of this type of spatially distributed array configuration. Figure 9.9(a) illustrates a hexagonal array and Figure 9.9(b) shows an example application. This makes it possible to detect the edges of objects. Information is provided about both the direction and speed of penetration in a defined detection range. A practical example

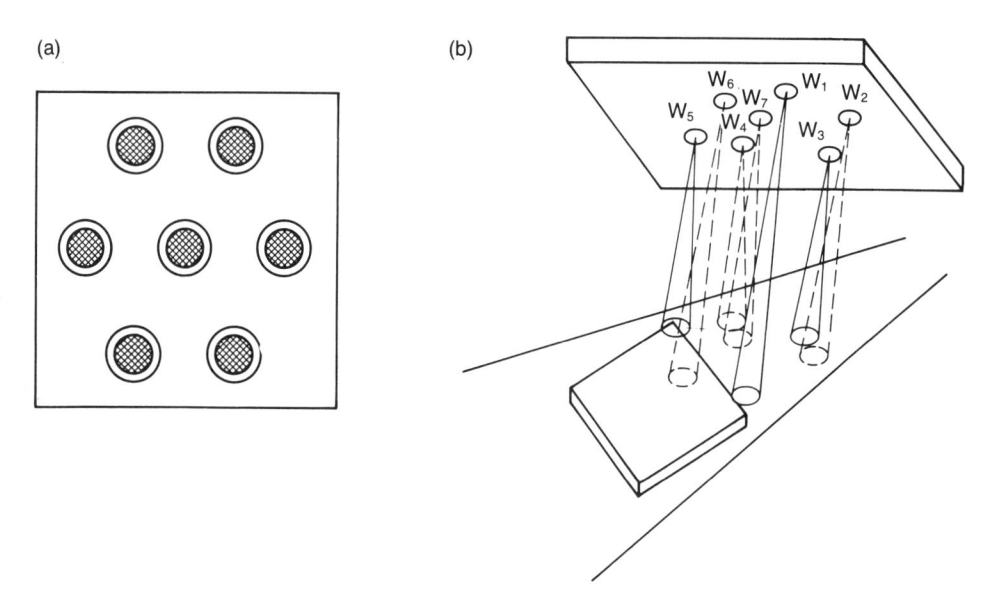

Figure 9.9 (a) Ultrasound transducer configuration in the form of a hexagonal array; (b) example application [326].

of the use of this type of sensor head is in connection with spray robots which are used to paint large, relatively flat surfaces (such as aircraft wings). In many applications it is necessary not only to analyze the geometry of the moving object but also to classify this object. An example of this sort of application is the supply of individual parts to an assembly robot. The robot must be able to recognize exactly which component has been passed to it for assembly. This can be achieved using linear arrays which detect the height/vertical profile of the object. Classification is then performed on the basis of a comparison of characteristic properties of the object and a pattern. The cross-correlation

$$\varphi = \sum_{l=1}^{N_W} [H_P(l) \cdot H_0(l)] \left(\sum_{l=1}^{N_W} H_P(l) \cdot \sum_{l=1}^{N_W} H_0(l) \right)^{-1/2} \tag{9.6}$$

where H_P is the height of the prespecified pattern, H_0 is the height of the object and N_W is the number of echoes, represents one classification feature. This expression is, of course, evaluated using a PC.

If no attempt is made to gather spatial information and only a single coordinate is considered, then a single-sensor classification can be performed which is much simpler than the array technique. High-resolution, wide-band electrostatic transducers are used for such applications. Figure 9.10 illustrates the structure of this type of configuration. Figure 9.11 presents a simplified example of the comparison of a pattern profile (A_P) with the echo from an object for classification (A_0). The comparison provides the number of overlaps. The correlation criterion is as follows:

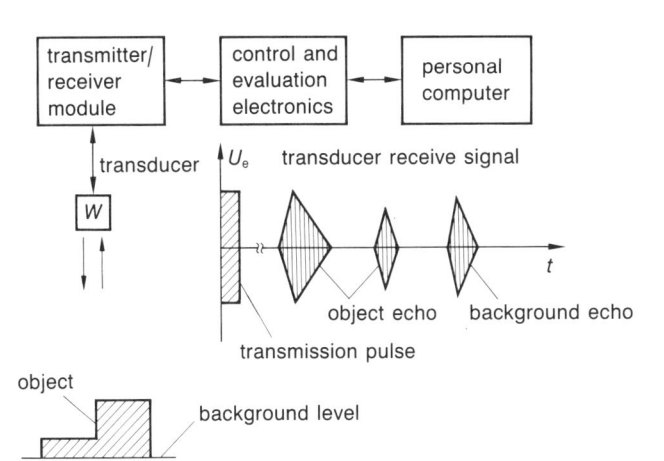

$$\varphi = \Sigma \, A_0 A_P / \sqrt{Z_0 Z_P} \tag{9.7}$$

Figure 9.10 Structure of a single-sensor classifier [326].

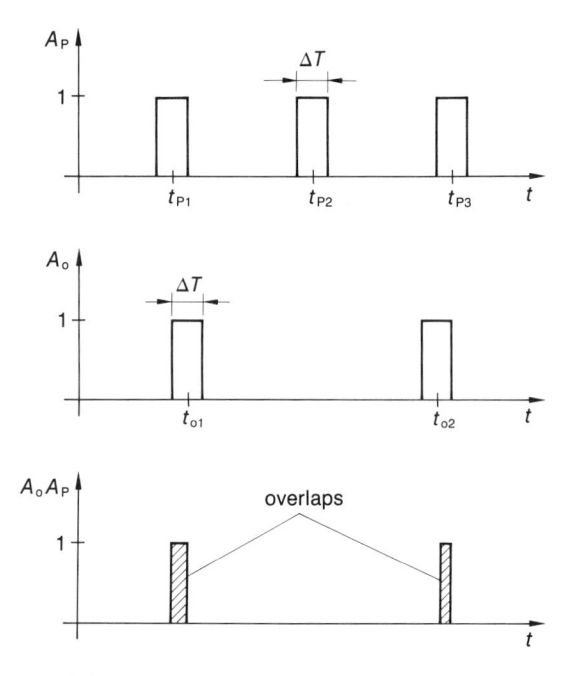

166 Sensors based on 'classical' measuring elements

Figure 9.11 Determining the correlation between the height profiles of object and pattern (pattern echo number $N_P = 3$, object echo number $N_o = 2$, number of overlaps $\Sigma A_o A_P = 2$).

The value of φ in the example illustrated in Figure 9.11 is 0.82. The object is allocated to the pattern for which the value of φ is greatest, provided that this exceeds a specified threshold. If this lower limit is not satisfied the object is classified as 'unknown'. Further applications for this type of ultrasound sensor include intelligent monitoring systems, code marking readers and, of increasing importance, applications in the field of driverless transport systems, for example for the transport of CIM objects. Ultrasound sensors are required for many throughflow measurement procedures [245]. A new application is in ultrasound tomographers for the examination of multiphase flows [327]. Ultrasound sensors which detect acoustic emissions are also of importance. These provide information about all types of breaks and fractures. A new area of application for ultrasound sensors is in the very broad area of process measurement. If we limit ourselves to the observation of fluid systems we can distinguish between two extremes in sound propagation, namely between propagation in simple liquids and propagation in complex systems consisting of more than one material. For the first the following apply:

$$c^2 = 1/\rho\beta \tag{9.8a}$$

$$\alpha = \omega^2/\rho c^3 (\eta_B + \tfrac{4}{3}\eta_S) \tag{9.8b}$$

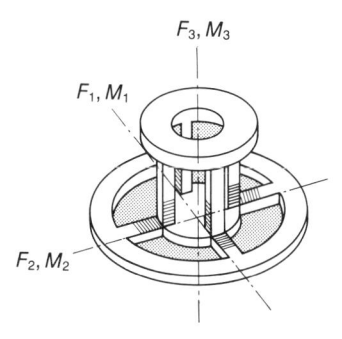

Figure 9.12 Structure of a force–moment sensor.

where β is the compressibility, ρ the density, and η_B and η_S are the bulk and shear viscosities, respectively. It can be seen that sound parameters have a defined relationship with the physical characteristic quantities of the medium which is being examined. If it proves possible to determine the relationships between the measured parameters c and α and these properties of the system, then it will be possible to provide interesting new information, on-line among other things, about complex material systems, the progress of reactions and process instabilities [328, 329]. By connecting other sensors it will then be possible to construct sensor systems which will open up a wide range of new applications in chemistry, the pharmaceutical industry, food industry and biotechnology.

9.4 Other principles

Interesting sensor principles can be realized by combining familiar procedures. Thus a combination of inductive and eddy current procedures has been used to create a sensor which measures the speed of rotation of ferromagnetic and non-ferromagnetic materials. Measurement is possible even if the rotating part exhibits a high degree of eccentricity [330]. Eddy current sensors on their own are suitable for detecting seams and fissures in ferrous containers. Various types of thermal conductivity sensor can be used to determine throughflow volumes. Correlation sensors are gaining in interest with the availability of fast FFT processors. In this case, the sensor principle may be based on ultrasound, induction or capacitance. Correlation sensors can be used to determine, among other things, throughflows, surface quality or surface speeds.

Intelligent robots require force–moment sensors which can be used to advantage in robot arm articulations. An example of this is presented in Figure 9.12. When the sensor is triggered, the WSGs which are attached at a number of points are stretched or compressed. This changes the electric resistance of the (eight pairs of) WSGs. The resistance values are converted into the forces and moments which are acting on the sensor.

10 New sensor materials

Advances in sensor technology are inconceivable without further development of the production technologies and improved or new materials. In the case of chemical sensors in particular, it has been shown how important knowledge of the materials is for a proper understanding of the detection mechanism. Only when the materials are understood can the sensor be optimized. At the moment no spectacular new sensor materials or principles are in prospect. Despite this, the coupling of materials development and sensor design technologies is stimulating continuous development. Table 10.1 provides a summary of new and existing materials for sensor technology. Although it does not claim to be complete, it nevertheless documents the variety of materials that can be used to achieve sensor effects.

Separate mention should be made of superconductors because of their significance as a new material and of adamantine structures because of their special properties. So far the only sensors to be based on superconductors are bolometers, in SQID devices (for the measurement of small magnetic fields, for example in brain currents) or in radiation detectors based on the Josephson effect [332]. There is currently no sign of any other developments.

Of great interest for sensor technology are polycrystalline thin diamond films, for example those produced using PECVD [333]. Their unique physical and chemical properties make them suitable for a wide variety of applications. For example, their extremely low chemical reactivity recommends them for use in highly corrosive environments. These films are extremely hard and have a very high Young modulus. They are well suited to applications where friction effects can be expected. Even SiC is more susceptible to friction than these films. Their thermal conductivity at room temperature is very high, some 2.5 to 5 times as high as copper. At the same time, they are very good electrical insulators. Fortunately diamond adheres excellently to silicon. This fact makes it possible to produce completely new sensor structures, for example by using micromechanical techniques, while simultaneously exploiting the excellent properties of diamond films [334].

Superconductors and diamond films have been mentioned separately as they may well fuel the most spectacular new developments of the future. At the same time, it should not be forgotten that a-Si:H, poly-Si and GaAs, which already have specific uses as optical fiber materials, also play an important role in the

Table 10.1 Summary of possible sensor materials

1. Semiconductors
2. Oxides and nitrides, doped and undoped
3. Optimized catalyst systems
4. Solid ion conductors
5. Ceramics and glasses
6. (Metallic) organic compounds and polymers
7. Membranes
8. Enzyme systems, antibodies, receptors, organelles, micro-organisms, animal and plant cells

---

development of new or the improvement of existing sensor technologies, while most recently poly-Ge has been used [335].

It is not easy to say how much further polymer materials will develop as materials for physical and chemical sensors. There is no shortage of proposals for mechanical, thermal, electro-optical, chemical and fiber optic sensors [336, 337]. A discussion of all the materials listed in Table 10.1 would need another book all to itself. For this reason I shall select just two classes of material which are of particular significance for sensor applications and discuss them in more detail below, namely PVDF and its modifications and the amorphous magnetics.

10.1 Piezoelectric polymers

10.1.1 Fabrication and properties

In 1969 the piezoelectric and pyroelectric properties of polyvinylidene fluoride (PVDF) were discovered. Pyroelectricity refers to the occurrence of charges as a result of temperature changes. Other organic substances were also later shown to be piezoelectric. However, none of these was comparable to PVDF in its piezoelectric activity. As a result, much research in the 1970s was devoted to using this property of PVDF. This development was further stimulated by the discovery that copolymers of difluoroethane (VF_2) with vinyl fluoride (VF), trifluoroethylene or TrFE (VF_3) or tetrafluoroethylene (VF_4) possess even better piezoelectric properties than PVDF. Thus this material is currently one of the most 'innovative' of sensor materials.

The piezoelectric and pyroelectric properties of PVDF come both from the property of the polymer itself and from the manufacturing conditions. PVDF is a semi-crystalline polymer (50% crystalline, 50% amorphous). The monomer unit, $VF_2CH_2CF_2$, has a large dipole moment. Head-to-tail polymerization of this monomer unit also gives the polymer a large net dipole moment. Despite this, further treatment of the polymer is necessary before it acquires the required properties. This is achieved by stretching the PVDF which has crystallized from the

170 New sensor materials

molten form until it is 4–5 times its original size. In this stretched PVDF film, the CF_2 dipoles still have no uniform orientation. It is only in a second step, the polarization of the films in electric fields of approximately 200 MV/m, that it becomes possible to achieve a preferred orientation of the dipoles along a line perpendicular to the film and thus obtain a stable macroscopic polarization. The PVDF is then piezoelectric. Figure 10.1 illustrates this process schematically. Since piezoelectric materials are anisotropic, the mechanical or electrical response characteristic is direction-dependent. The numbers in the upper left-hand corner identify the axes and define a convention which makes comparisons possible: 1 corresponds to length, 2 to width and 3 to thickness. Two indices are used to designate constants. The first identifies the axis of the electric field, the second the axis of the induced mechanical distortion or applied voltage. The electromagnetic behaviour of such a PVDF film is described by the piezoelectric constant d. This characterizes the way the film geometry changes when an electric field is applied. Thus:

$$d_{3i} = \frac{\text{deformation along axis } i \text{ (m/m)}}{\text{electric field applied along axis 3 (V/m)}} \tag{10.1}$$

Conversely, when an external mechanical force acts on the piezoelectric film, the film is deformed. This results in a change in the surface thickness of the material. In a static configuration it would be necessary to store these charges. In reality, there is a charge flow with the time constant RC (R is the resistance of the piezoelectric

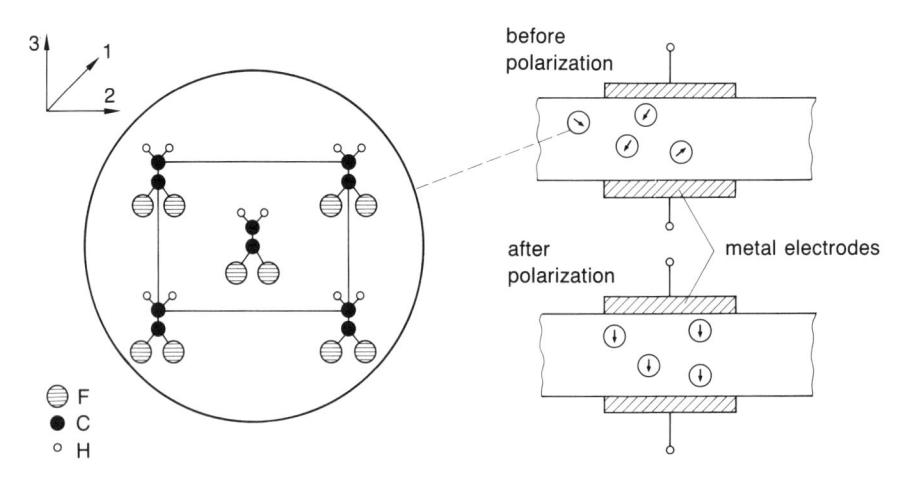

Figure 10.1 Figure 10.1 Diagram of field-induced dipole alignment in PVDF. The polymer molecules are in the polar β modification [342].

film and C is its capacitance). In consequence, it is only possible to detect dynamic processes. This is described by the constant g. For example,

$$g_{3i} = \frac{\text{field applied along axis 3 (V/m)}}{\text{strain acting along axis } i \text{ (N/m}^2\text{)}} \tag{10.2}$$

Two numerical examples will be presented here to provide a rough idea of actual values. A force of 4 N is to act on a piezoelectric film with an area of 2×2 cm^2 and thickness 110 μm. In the case of compression alone the resulting voltage U is described by

$$U = Et = g_{33}\sigma t \tag{10.3}$$

where σ is the compression pressure and t is the thickness of the film. Given g_{33} from Table 10.2, the value of U is calculated to be approximately 0.4 V. If the same force is applied in the 31 plane the voltage increases to approximately 40 V. Similar calculations could be performed for the pyroelectric effect. Given ΔT of a few degrees, the effect is also of a few volts. This means that considerable effects can be achieved when PVDF is used as the sensor material. PVDF is commercially available in thicknesses of 9–800 μm at widths of 450 μm and lengths of up to a few kilometres. The PVDF film is flexible, extremely light and has a plastic toughness. It can be used in temperatures between $-40°C$ and $100°C$. Table 10.2 lists the most important parameters [339, 340]. In Table 10.3 PVDF is compared with other piezoelectric materials. From this comparison it is possible to draw important conclusions concerning the advantages and disadvantages of this material in sensor applications.

Piezoelectric films can function in a frequency range of 1 Hz to 100 MHz. This wide bandwidth means they can be used in applications ranging from breath control to high-frequency ultrasonic environments. They possess a wide dynamic range. Their acoustic impedance level is low and similar to that of water. This is of extreme importance when the material is used as a hydrophone or in medical ultrasound

Table 10.2 Typical properties of piezoelectric films

d_{31}	piezoelectric strain constant	23×10^{-12} m/V
d_{33}		-33×10^{-10} m/V
g_{31}	piezoelectric voltage constant	216×10^{-3} V m/N
g_{33}		-339×10^{-3} V m/N
k_{31}	electromagnetic coupling factor	12% (1 kHz)
C	capacitance	380 pF/cm^2 for 28 μm film
p	pyroelectric coefficient	-35×10^{-6} C/m^2
	operating temperature range	$-40°C$ to $100°C$

Table 10.3 Comparison of different piezoelectric materials

Property	Unit	PVDF	PZT	BaTiO$_3$
Density	10^3 kg/m^3	1.78	7.5	5.7
Relative DC	ε_r	12	1200	1700
d_{31}	10^{-12} C/N	23	110	78
g_{31}	10^{-3} V m/N	216	10	5
k_{31}	at 1 kHz	12	30	21
Acoustic impedance	10^6 kg/m^2 s	2.7	30	30

technology and NDT. Piezoelectric films have a high degree of elastic resilience and consequently a low mechanical quality factor Q. The electric signals are only slightly distorted when converted into mechanical signals. Its low DC means that PVDF can be subjected to stronger electric fields than piezoceramics, thus compensating for the disadvantages caused by the low piezoelectric coupling factor. PVDF is resistant to most chemical substances and is not sensitive to radiation. It is easy to cut and can be shaped to produce any required geometry or laminate. Electrodes can be attached using standard methods.

Even better properties, that is to say, increased piezoelectric activity and a better electromagnetic coupling factor, can be achieved using copolymers, especially of type P (VF$_2$TrFE). Research into these materials is under way and new possibilities can be expected for the future [321, 340].

Of course, piezoelectric films also have drawbacks. Among these are their very limited temperature range as well as problems of long-term stability caused by the ageing characteristics of polymers. The flexibility of the film means that high-power components cannot be produced.

10.1.2 Applications

In the first applications, piezoelectric films were used as a replacement for piezoceramics. Since then the range of applications has grown and now includes the fields of quantitative sensors, robotics, acoustic and optical devices, computers and office automation, medical equipment, military applications, the automobile industry, sport and leisure goods [336, 341, 342]. Some examples are listed below.

Quantitative sensors include vibration, acceleration, power, temperature, infrared and ultrasound sensors. Piezoelectric films fill the gap between, for example, WSGs and classic acceleration sensors very well. Like the latter, they are mounted on motors or machine parts and have a high degree of sensitivity to, for example, changes in bearing friction or harmful vibrations [339]. Temperature sensors make use of changes in temperature caused by the absorption of incident infrared radiation. Very thin PVDF films, for example of a thickness of 9 μm, are used to achieve fast response times. Thin gold or polymer layers are applied to the PVDF film and act as infrared-absorbent materials. The piezoelectric effect, which occurs at the same time as the pyroelectric effect and vice versa, has to be minimized in such sensors. Figure 10.2 illustrates a cylindrically arranged pressure sensor [342]. When the pressure in the tube changes, the tube is subjected to a radial strain. This causes tangential stresses in the film and these can be detected. Medical applications, such as pressure measurements in arteries, as well as industrial flow measurements in tubes are conceivable. PVDF-based ultrasound sensors for object recognition, distance measurements and non-destructive materials testing with quantitative evaluations of materials faults also belong to this category. PVDF is a very interesting material for high-resolution tactile sensors in robotics applications. Figure 10.3 illustrates a sensitive piezoelectric cell consisting of a PVDF film and electrodes. When a number of such cells are grouped together to form an array and are connected to information processing electronics an area-sensitive tactile sensor is formed. Researchers are also attempting to develop PVDF-based touch-sensitive sensors [343] to provide information for object characterization. Research into the development of artificial skin is taking a similar direction [344].

PVDF has led to many developments in the hydrophone field. Using point-polarized films, it is possible to measure ultrasound fields to a high degree of accuracy. This is very important for medical applications. Hydrophones are also of significance for sonar technology, that is to say, for underwater location and orientation, in both the military and civilian spheres.

PVDF is an interesting cladding material for optical fibers where it modulates the light which is transported in the fiber core. The pyroelectric and piezoelectric properties of PVDF are being increasingly employed in security technology. Thus the thermal radiation emitted by humans can signal their presence. Piezoelectric films in the floors of security zones are used to detect the presence of intruders. Vibration sensors are used in car alarm systems or in window panes.

The importance of PVDF as a sensor material in medical applications is growing. The use of ultrasound tomography is one example of an application in which the good acoustic coupling and the focusing properties of PVDF transducers are exploited through the use of selectively adjustable geometries. However, this

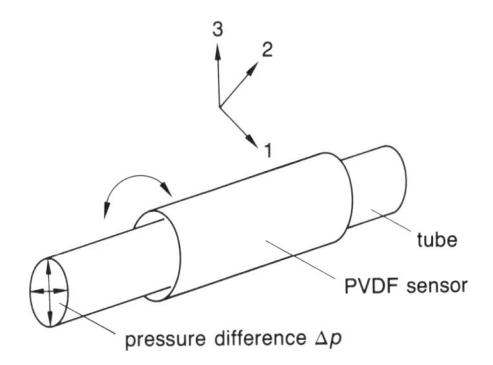

Figure 10.2 Cylindrical arrangement of a PVDF film for the detection of pressure changes in a tube [342].

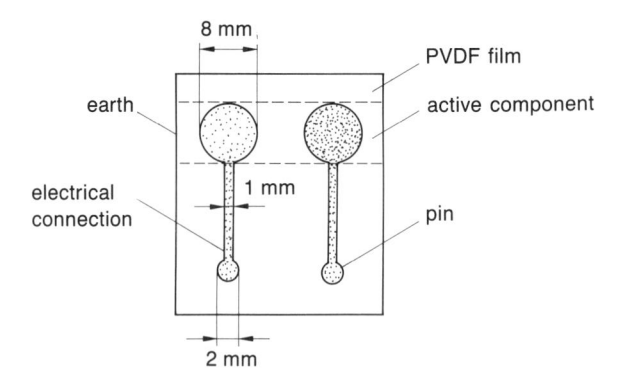

Figure 10.3 Tactile sensor cell.

material is also suitable for blood pressure measurements, eye examinations, artificial skin or for the thermal imaging of body surfaces.

The range of applications of PVDF sensors even extends to the sports, leisure and games sectors. Thus it is possible to use them in a boxer's gloves to measure his punching power. There are 'shoe sensors' for joggers. The 'intelligent' doll can tell she is being kissed by means of a PVDF pressure sensor and will respond to it.

There are now a large number of PVDF and copolymerizate manufacturers, especially in Japan. In Europe, the Belgian company Solvay produces a wide range of materials.

10.2 Amorphous metals

Amorphous metals, also known as metal glasses, form a new class of materials which has been a focus of interest for materials scientists and users for more than a decade. A number of features distinguish them from the crystalline metals [345]. Thus certain alloys possess excellent soft magnetic properties while at the same time exhibiting certain desirable mechanical properties such as a high degree of hardness or elastic limit. These properties and combinations of properties make these materials interesting for sensor applications. Apart from this, they also have the advantage that these properties can be selectively adjusted through the correct modification of the alloy composition. This makes a wide range of physical properties accessible. For example, it is possible to manufacture soft magnetic materials with either a high level of magnetostriction or with a level of magnetostriction approaching zero. The first group is an ideal material for magnetoelastic sensors, the second for magnetic field sensors. However, there are still other physical properties, such as the high field dependence of the modulus of elasticity and the electrical resistance of many alloys or a high k factor, which make

Amorphous metals 175

these materials attractive for sensor applications [346]. Table 10.4 collates the possibilities for the use of amorphous alloys in sensors. Possible materials for magnetoelastic sensors are, for example, $Fe_{80}Si_{13}B_7$, $Fe_{80}B_{14}Si_6$, $Fe_{40}Ni_{38}(Mo, Si, B)_{22}$, $Co_{75}S_{15}B_{10}$, while $(Co, Fe)_{70}(Mo, Si, B)_{30}$ is suitable for magnetic sensors. These materials are marketed by companies such as Vacuumschmelze (Hanau).

10.2.1 Magnetic field sensors

A well-known procedure for measuring constant magnetic fields makes use of the premagnetization of magnetically alternating core probes by the constant field that is to be measured. The change this causes in the characteristic curve of the ferromagnetic material is detected as a change in induction [312]. The higher the level of ferromagnetism in the material, the more prominent this effect is. Cobalt-rich amorphous alloys with magnetoresistive levels close to zero and a high degree of permeability are excellent materials for this type of sensor. Two variants will be presented [346].

One principle makes use of the effect of reversible permeability. Given reversible permeability μ_{rev}, the permeability of the alternating field with a low level of modulation and a superimposed constant field is described by

$$\mu_{rev} = (1/\mu_0)(\Delta B/\Delta H) = f(H_0) \qquad \text{as } \Delta H \to 0 \qquad (10.4)$$

This is illustrated in Figure 10.4(a). The inclination of the depicted lancet corresponds to μ_{rev}. The amorphous alloy exhibits a particularly high degree of sensitivity to the constant field (Figure 10.4(b)). Strip- or rod-shaped cores made of this material, surrounded by a coil, are used as measuring probes. Figure 10.5 presents the circuitry of this type of magnetic field sensor and the characteristic curve that it yields.

Table 10.4 The use of amorphous alloys in sensors

Magnetic sensor $L \sim \mu_{rev}(H_0)$	Saturation core probe
Magnetoelastic effect $\mu = \mu(\sigma)$	Pressure sensor
	Force sensor
	Torque
	Impact sensor
Magnetoresistive effect $R = R(H)$	Magnetic field sensor
	Reading head
Body resistance effect $R = R(L)$	Force sensor
Stepped magnetic reversal effect $M = M(H_{crit})$	Magnetic field sensor $E = E(H)$
Elasticity-effect	Temperature sensor $E = E(T)$

176 New sensor materials

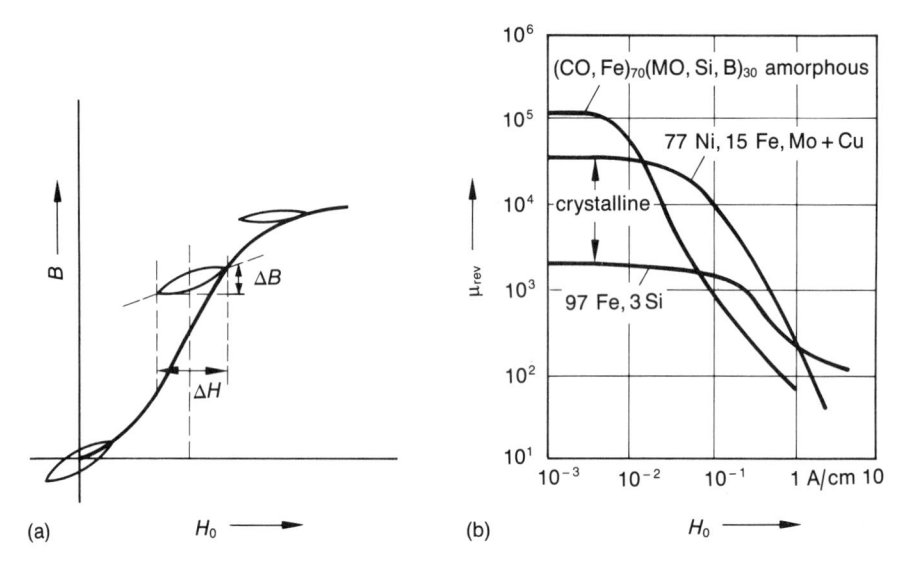

Figure 10.4 Reversible permeability: (a) definition; (b) measured values for an amorphous and a crystalline material [346].

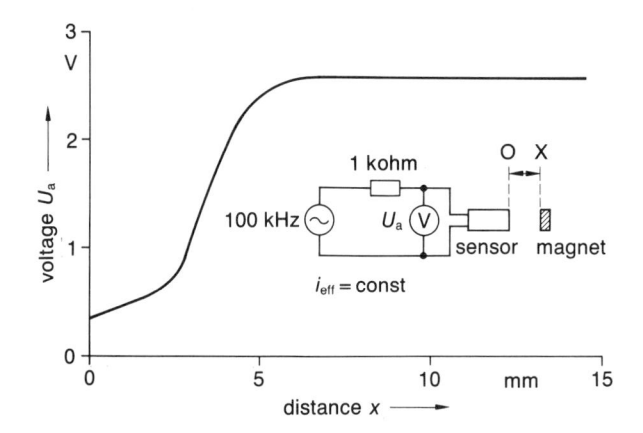

Figure 10.5 Magnetic field probe as proximity switch [346].

An alternative magnetic sensor is the *saturation core probe*. In this device the magnetic probe is moved fully into the material, that is to say, into the saturation zone. However, as yet there have been very few low-cost applications. This technology has long been applied in the form of the Förster probe as a highly sensitive procedure for measuring magnetic fields, for example for geophysical and astrophysical, military and navigational purposes [347]. Saturation core probes made of amorphous metals are opening up a new range of applications [348]. These consist of a sensor coil with a ferromagnetic core, whose permeability is affected by external magnetic fields, and the necessary evaluation electronics. Magnetostriction-free, Co-rich amorphous alloys with low coercive force (5 mA/cm), high relative permeability (static value between 5×10^4 and 5×10^5), a low saturation magnetization (0.5 T) and low eddy current and hysteresis losses are again a suitable core material. The induction can be controlled using any field, for example the earth's magnetic field, an electric field or a permanent magnet. The control magnets can be made, for example, of somarium-cobalt alloys. The high permeability of the core means that external fields as weak as 0.1–1 mT are sufficient to premagnetize it to the saturation range and thus to cause a dramatic fall in induction (Figure 10.6). In this way, a high response sensitivity is achieved. The shape of the curve in the transitional range depends on the form of the hysteresis loop and the amplitude of the alternating current which is used for the measurement. It can thus be adapted to meet the needs of the particular application (sharp transition for position sensors, flat transition for linear magnetic field, distance and rotational speed sensors). Carrier frequency measuring bridges, magnetically controlled oscillators and pulse–height magnetometers are used to evaluate the response, and sensitivities of up to 10 V/mT are achieved. The first applications for these sensors have been in the automobile industry. Figure 10.7 illustrates a speed-of-rotation sensor with a saturation core probe.

Saturation core probes can also be used to determine the time and duration of injection in diesel injection nozzles, as electronic compasses in cars (very important for the traffic control systems of the future) and to count vehicle numbers. They are more sensitive than Hall sensors or magnetostrictive sensors by several orders of magnitude. However, the electronics required for analysis are more complex.

A particular area of application for saturation core probes is in security technology, for example in shop anti-theft systems [346] or to prevent the theft of

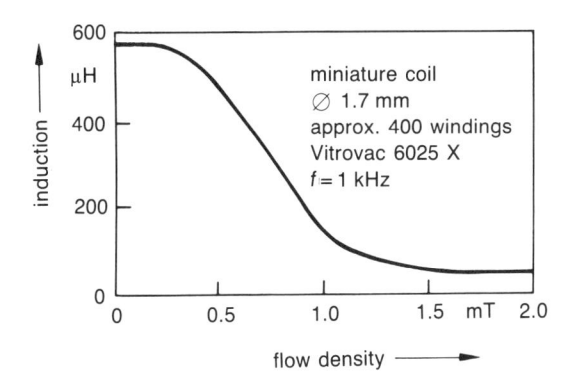

Figure 10.6 Characteristic of a sensor coil. The fall in induction is caused by displacing the operating point on the magnetization curve [348].

178 New sensor materials

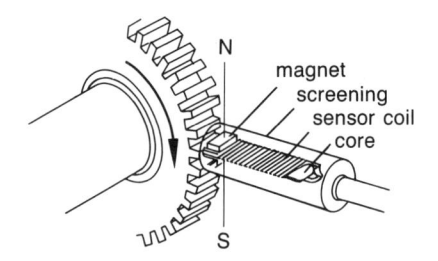

Figure 10.7 Speed-of-rotation sensor with saturation core probe for recording extremely slow rotation speeds [348].

books from libraries [349]. The sensor element consists of a strip of highly permeable material which is attached to the item to be protected and is detected in a magnetic field. The magnetic coil system consists of an excitation winding and a detection winding (Figure 10.8). The field of the excitation winding W_1 drives the security strip to saturation and generates magnetic reversal effects in it which are detected at the detection winding W_2. The fundamental frequency can be suppressed, making it possible to evaluate only certain harmonics. These generally lie between the second and twentieth harmonic and may extend to the 100 kHz range. What is crucial is that while these harmonics can be generated even at low excitation field values they are still clearly differentiated from other signals. This means that the security loop should be highly permeable, aligned at right angles to the hysteresis loop and be as insensitive as possible to mechanical influences. These conditions are largely satisfied by Co-rich amorphous alloys.

10.2.2 Magnetoelastic sensors

Sensors which employ the magnetostrictive effect have long been used as force and torque sensors. In these sensors a change of length or strain caused by external

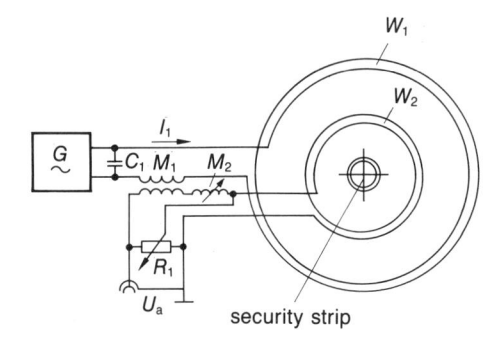

Figure 10.8 Coil configuration and circuitry for anti-theft system [346].

forces has the effect of changing the magnetic permeability or distorting the hysteresis loop. Although these magnetoelastic effects occur in all ferromagnetic alloys, they are particularly prominent in the amorphous magnetics and it is only in these that the effect can be used. The fundamental principles of magnetoelastic sensors are shown in Figure 10.9. While pressure membranes are now familiar items, it only became possible to realize the rod- and ring-shaped sensors when amorphous materials became available. Force and strain sensors have been developed and make use, for example, of ring-shaped cores [351]. Alongside the shape and the material, the electronic evaluation circuitry is also a decisive factor in magnetoelastic sensors. Some possibilities are presented in Figure 10.10. Of interest is the fact that either frequency-analog or binary output signals can be obtained. Using this principle it is possible to develop position indicators which function at a resolution of the order of 0.1 mm.

The use of amorphous metals in torque sensors has opened up a completely new range of possibilities [352–354]. Torque is an important measurand in both static and rotating shafts. It is also necessary to know the value of the torque for the monitoring and evaluation of the operating processes of the connected machines (for example, machine-tools). For many of these tasks there are already tried and tested measuring systems, especially those using WSGs. However, their disadvantage lies in the complexity of the signal transmission from the rotating shaft to the fixed housing. For this reason, contact-free systems, using inductive or telemetric techniques [360], are increasingly finding favour. Some of these require the transmission element to be mounted on the shaft. Configurations using

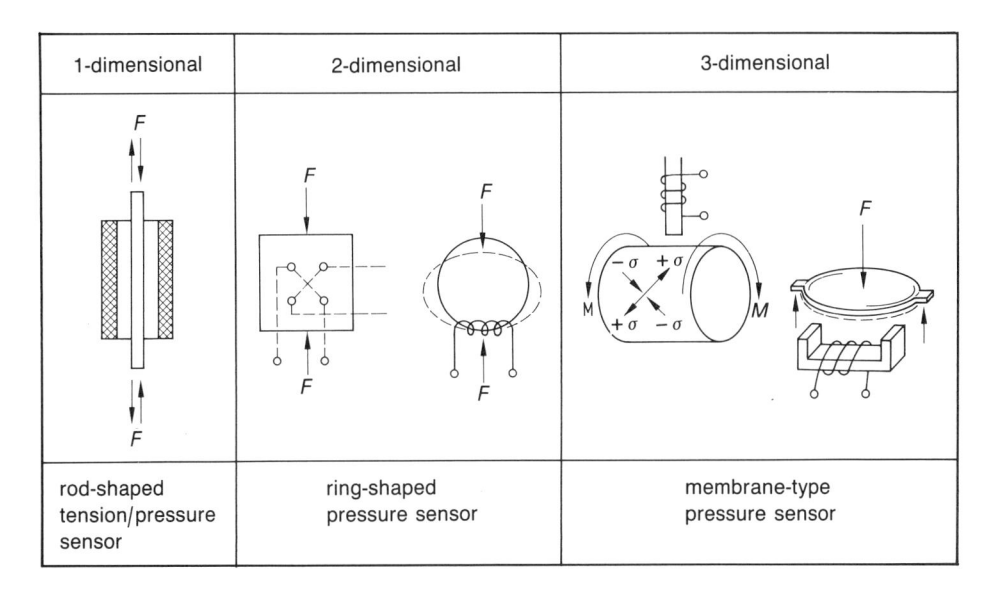

1-dimensional	2-dimensional	3-dimensional
rod-shaped tension/pressure sensor	ring-shaped pressure sensor	membrane-type pressure sensor

Figure 10.9 Fundamental principles of magnetoelastic sensors [346].

180 New sensor materials

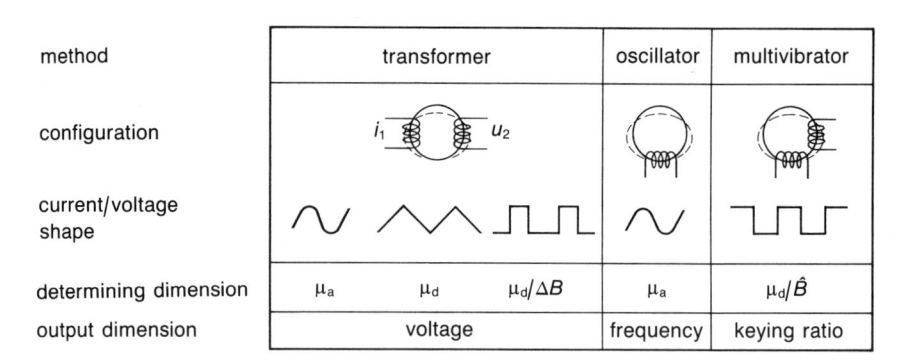

method	transformer			oscillator	multivibrator
configuration	i_1 ⬮ ⬮ u_2			◯	◯
current/voltage shape	∿	⋀⋀	⊓⊔⊓	∿	⊓⊔⊓
determining dimension	μ_a	μ_d	$\mu_d/\Delta B$	μ_a	$\mu_d/\hat{B}$
output dimension	voltage			frequency	keying ratio

Figure 10.10 Methods of evaluation for ring sensors.

amorphous metals are simpler. In these, amorphous bands, films [352] or strips of a particular geometry [353] are glued or welded to the shaft at which the torque measurement is to be performed. The bands or films are pre-tensed (with moment M_0) to a defined value so that there is an anisotropy, which corresponds to the torsion strain field, in the amorphous layer at 45° to the axis of the shaft. This can be described by the constant of anisotropy

$$K = 48\lambda_S M_0 G_A / \pi d^3 G_{SH} \tag{10.5}$$

where λ_S is the saturation magnetostriction, d is the diameter of the shaft and G_{SH} and G_A are the shear moduli of the shaft and the amorphous band, respectively. Figure 10.11(a) illustrates the principle. When the measuring moment M is applied, the constants of the two bands change:

$$k_1 = k'(M_0 + M)$$
$$k_2 = k'(M_0 - M) \tag{10.6}$$

A change in the permeability between the bands occurs and yields the characteristic curve illustrated in Figure 10.11(b). The technique can be made more attractive if amorphous layers are deposited directly on the surface of the shaft. The problems associated with the connection methods, such as gluing, are then avoided.

10.2.3 Other sensor principles

The modulus of elasticity of tempered amorphous alloys with the composition $Fe_{40}Ni_{38}Mo_4B_{18}$ differs greatly between the magnetized and nonmagnetized states. This ΔE effect can amount to as much as 20% and is useful in delay lines or

Amorphous metals 181

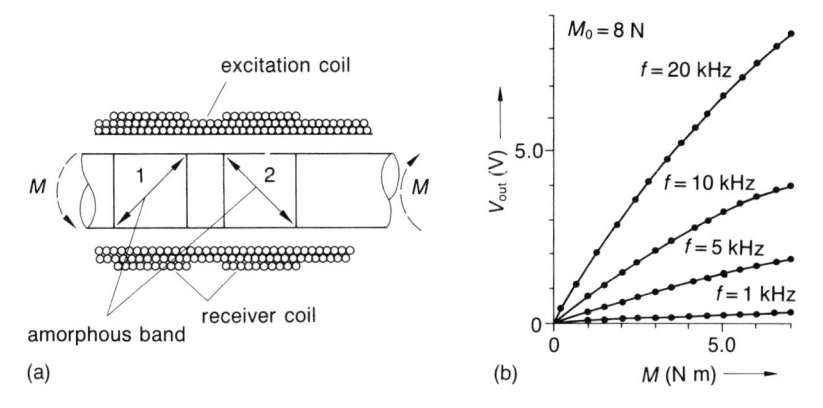

Figure 10.11 Torque sensor [352]: (a) with film-coated shaft; (b) characteristic curve.

retarding devices (Figure 10.12). The magnetic field strength is measured through the change in the sound speed c or the change in the travel time

$$\Delta c \sim \sqrt{\Delta E/\rho} = f(H) \tag{10.7}$$

Devices based on this principle can also be used as temperature sensors if the material's existing dependence on $E(T)$ is used [356].

The magnetoresistive effect refers to the change in the electrical resistance of ferromagnetic alloys under the influence of a longitudinal or transverse magnetic field. Depending on the material used, the relative change in resistance $\Delta R/R$ can vary from a few parts per thousand to several per cent. In sensors this effect can be used for the localized detection of static and dynamic magnetic fields.

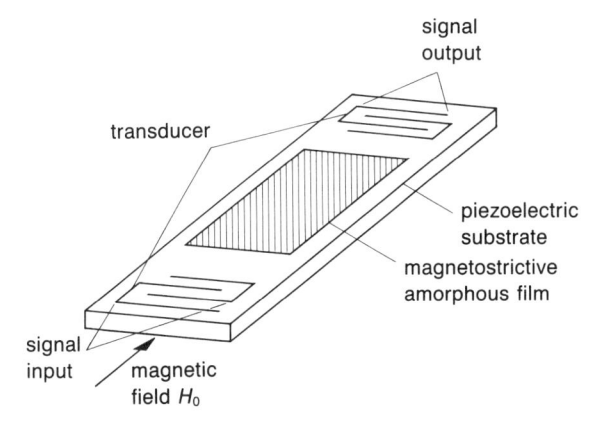

Figure 10.12 Magnetically regulated delay line [346].

Among the amorphous alloys, it is possible to find some with a k factor of up to 4. These include alloys with the composition $Fe_{57.5}Cr_{30}Si_{12.5}$. Since suitable tempering gives this material a temperature coefficient equal to zero over a wide range, it is of considerable use as a WSG.

10.3 The Wiegand sensor

Specially manufactured wires made from ferromagnetic material exhibit a spontaneous change in the direction of the magnetic field when a certain field strength is exceeded. This effect has long been known as the *Large–Barkhausen discontinuity*. However, it was not until the mid-1970s, after a suitable technology for the manufacture of these wires had been discovered by Wiegand [357], that the effect was used. A Wiegand wire is made of vicalloy, an alloy composed of cobalt, iron and vanadium whose mechanical and magnetic hardness can be increased by cold shaping. A soft magnetic wire with a diameter of approximately 0.25 mm is cold-shaped by a process of alternating opposite twisting and is then stretched slightly. The properties which this lends the wire are then stabilized in a heat treatment stage. During this process the deformation of the thin cladding of the wire is greater than that of the core. In this way a magnetically and mechanically hard shell with a ferromagnetic core is produced. The deformation profile along the wire is very pronounced. As a result, the direction of magnetization of the Weiss zones is parallel to the axis of the wire. The actual sensor effect can now be explained with reference to Figure 10.13 [358]. If we assume that the core and the shell are magnetized to saturation in the same direction by a field of 100 A/cm (this is known as the *asymmetrical operating mode* of the sensor), then a relatively weak, increasing, opposing field of approximately 15 A/cm (reset field strength) causes a spontaneous reversal of the magnetization of the core. The associated change in flux generates a pulse in a coil wound around the sensor wire. This pulse is known as the *reset* pulse. If a strong external field now acts in the original direction, the magnetization of the core reverses to its original direction at the trigger field strength. As in this case the effect of the external field is supported by that of the shell, the magnetic coupling of the shell and the core causes a high speed of flux change. A pulse with a strength of several volts can be detected in the coil with a half-power width of some 20 µm. The great advantage of the Wiegand sensor over inductive sensors lies in the fact that the pulse amplitude is largely independent of the speed of change of the magnetic field. Moreover, it requires no external power supply and operates over a wide temperature range of $-200°C$ to $+200°C$. It is very robust and is insensitive to interference. Recently the usefulness of Wiegand wires has been extended by the use of amorphous metals. It has been possible to demonstrate in the laboratory that heat treatment of an Fe-rich magnetostrictive band, for example $Fe_{80}B_{20}$ or $Fe_{81}B_{17}Si_2$, generates different mechanically pre-tensed layers in it [359]. Material prepared in this way has properties similar to

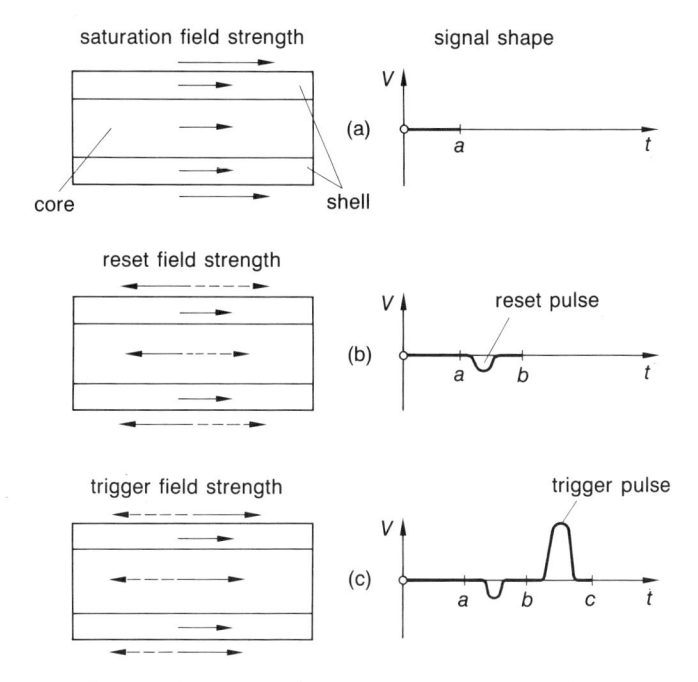

Figure 10.13 Influence of an external field on the core and shell of a Wiegand wire [358].

Wiegand wires. However, it has the advantage that the reversal of the Weiss zones requires only about one-fiftieth of the magnetic field strength that is needed in conventional Wiegand wires.

Wiegand sensors have been commercially available since the early 1980s. They have a modular construction and consist of a wire, a sensor coil and a mechanism for changing the magnetic field. The magnetic field can be provided by the field of an electromagnet which changes with time. Another possibility is the relative movement of a permanent magnet and the Wiegand wire. This type of sensor is well suited for the measurement of incremental distances, speeds of rotation or angles, as well as for the measurement of the speed of angular or linear movements. Figure 10.14 gives an example of a high-resolution speed-of-rotation sensor. The sensor wires are affixed at uniform intervals around the circumference of a rotationally symmetrical, non-ferromagnetic body and are parallel to its axis. Thanks to the symmetrical distribution of the field, the configuration can be altered to run in the other direction, in which case the polarity of the sensor coil is reversed. This type of sensor is used in vehicle construction.

Figure 10.15 shows an incremental linear transducer. The indicator bar contains grooves at intervals t in which Wiegand wires are located. The moving reading head registers the trigger pulses. When the direction of head travel reverses,

184 New sensor materials

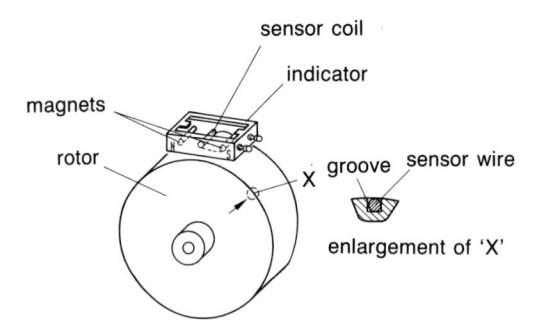

Figure 10.14 Speed-of-rotation sensor with rotor and indicator.

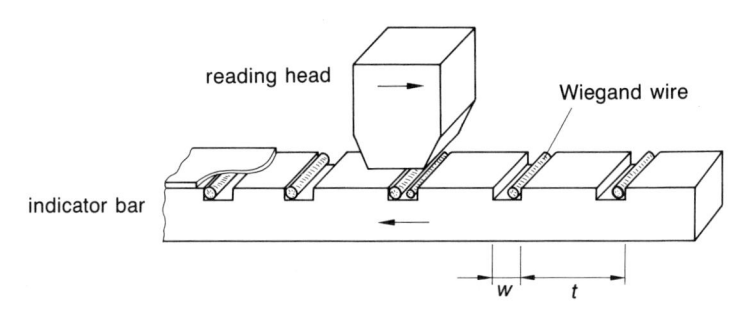

Figure 10.15 Incrementally functioning linear transducer with Wiegand wires.

a position hysteresis occurs which is proportional to *w*. This sensor is very robust, provides a moderate, localized resolution and can therefore be used in large-scale engineering applications, in hydraulic controls, etc.

Wiegand wires can also be used in coded card applications. For example, two rows of vertically aligned Wiegand wires can be located in a plastic card which is moved over a reading head. The reading head contains the sensor coil which is subjected to a permanent magnetic field. The Wiegand wires pass through two separate air gaps and their direction of magnetization changes in the magnetic field. This generates a characteristic stream of pulses in the sensor coil. These pulses have differing polarizations which can be used for identification purposes.

11 Resonance sensors

Most sensors used today are analog sensors. Compatibility with information processing microelectronics is achieved through an analog–digital conversion process which is more or less complex depending on the sensor techniques employed. The ideal would be a sensor which was integrated with the processing electronics and which provided a microprocessor-compatible, bus-ready output signal. Between these two extremes we find a useful and necessary step in the design and production of sensors with digital or frequency-analog outputs [361]. There are still few examples of ideal digital sensors: the flip-flop sensor (Section 4.5.2) is one; in principle, all length or angle decoders can also be counted among their number. The latter provide a four-channel digital signal from the optoelectronic scanning of coded marks on code discs [313]. They have long been known and do not form part of the subject matter of this book. Instead, it is the frequency-analog sensors on which I shall concentrate in the following discussion. These provide an output signal which is easy to digitize or process electronically. They are frequently very precise sensors and are thus of considerable interest to users. It is necessary to distinguish between two types of frequency-analog sensor:

- sensors based on electric oscillators in which the sensor is the frequency-determining element; and
- sensors with mechanical resonance structures.

The first category includes the ring oscillators, for example in the form of the smart pressure sensor (Section 4.5.2). Although attempts are being made to develop them [362], there are still very few examples of this type of sensor. The second category includes those in which the resonant frequency or the frequency distribution of a mechanical structure is measured. The quantity to be determined is then calculated from this measurement. These resonance sensors are rapidly gaining importance [363]. They are significant not only because of their miniaturizability but also because of the resonance effect. They cover a wide range, from the micromechanical resonance sensor, for example the vibration sensor (Section 4.8.1), to the bulky density or throughflow meters [368]. The following overview will therefore not be exhaustive but will instead simply present the key types of sensor and their applications: the quartz resonance sensor; the SAW sensor; and resonance sensors

186 Resonance sensors

for density, filling level and throughflow. Micromechanical sensors will not be described again although they in part belong to this sensor category.

11.1 Quartz resonance sensors

Crystalline quartz is a material with exceptionally stable elastic properties and with piezoelectric capabilities. Large quantities of it have been used for many years as the frequency standard in electrical circuitry, and our knowledge of it is therefore very exact. Using quartz resonators, it is possible to achieve mechanical quality factors of up to 10^5 and consequently also high resolutions. This is a great advantage of these sensors.

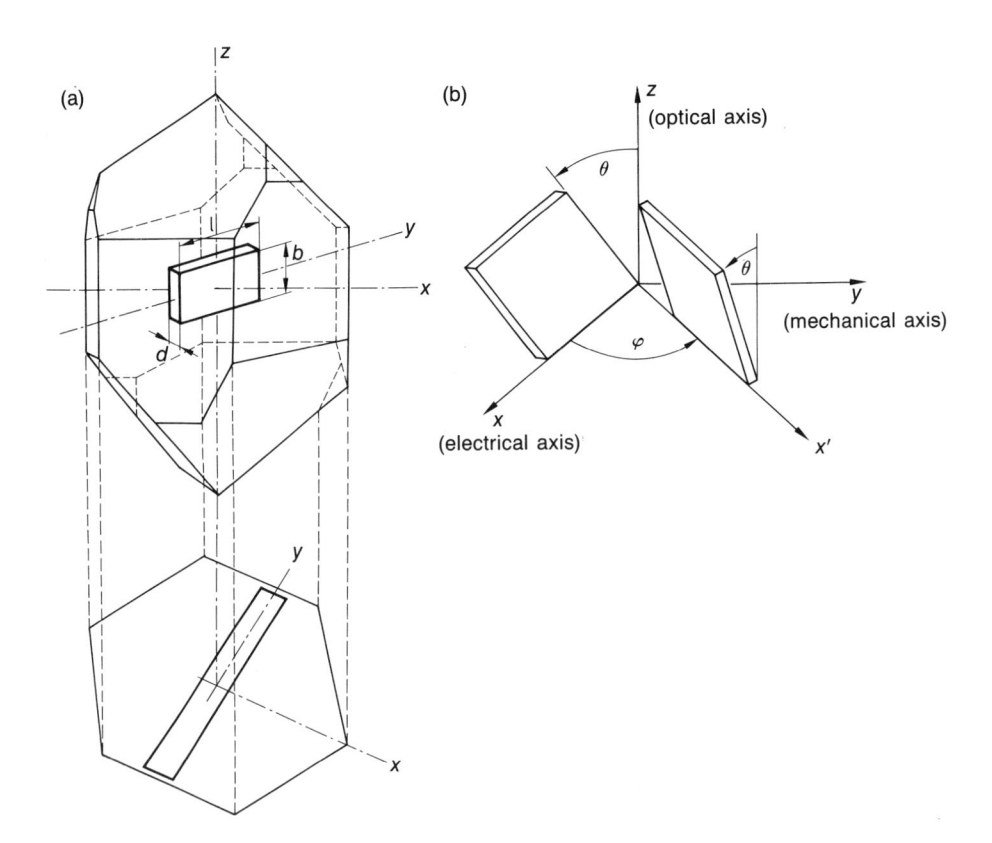

Figure 11.1 (a) Quartz monocrystal and orientation of a quartz oscillator (z: trigonal axis; x: transverse axis of oscillation; y: longitudinal axis of oscillation); (b) angles of cut θ and φ between the quartz oscillator and the optical (z) and electrical (x) axes.

Quartz resonators are thin lamellae of crystalline quartz on which electrodes are deposited. These lamellae are cut from quartz monocrystals (Figure 11.1). Depending on the crystallographic orientation of the cut piece and the electrode configuration, different types of resonator can be obtained. These include transverse shear, surface shear, transverse, torsional and bending oscillators [394]. Each type of oscillation is suited to certain sensor applications. The most frequently used is the transverse shear oscillator. Its resonant frequency is given by

$$f_r = (\tfrac{1}{2}d) \cdot c = (\tfrac{1}{2}d)\sqrt{c_{11}/\rho} \qquad (11.1)$$

where d is the thickness of the quartz lamella, ρ its density and c_{11} the coefficient of elasticity in the direction of the sound wave which is propagating at velocity c. Since the trigonal crystal structure of crystalline quartz gives it highly anisotropic properties, the resonant frequency is also dependent on the orientation of the crystal lamella with reference to the crystal axes (Figure 11.1(b)). As the coefficient of elasticity is temperature-dependent, the frequency is also temperature-dependent, and this temperature dependence changes with the angle of cut. Figure 11.2 shows an approximation for the dependence of the linear temperature coefficient α' as given by

$$f = f_0(1 + \alpha' T) \qquad (11.2)$$

as a function of the angle of cut between the plane of the wafer and the optical axis. In fact, the temperature dependence of the frequency is precisely described by a cubic parabola. Figure 11.2 nevertheless provides important information. At around

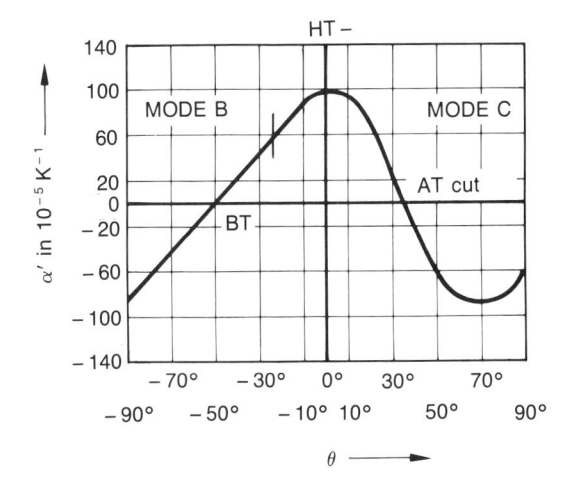

Figure 11.2 Dependence of the linear temperature coefficient on the angle of cut θ [376].

30°, the so-called AT cut, the temperature coefficient has the value zero. AT quartzes are transverse shear oscillators and are used in practically all technical quartz applications. Their fundamental wave covers a frequency range from 750 kHz to 20 MHz. Even at high frequencies their thickness is only a matter of fractions of a millimetre. This means that they can break easily. Frequencies higher than 20 MHz can be obtained using the harmonic excitation process. These AT quartzes are used with differing geometries as frequency standards. However, another interesting area of application for AT quartzes is as sensors. Their longest and best-known application in this field is for measuring the thickness of films in the vacuum deposition systems used by the optics and semiconductor industries. In these applications, a quartz oscillator which has been electrically excited to a resonant frequency, usually the fundamental wave, is coated with the substrate. The change in the frequency caused by the coating of the quartz is then used as a measure for the thickness of the applied layer. If we take as our basis the model presented in Figure 11.3, then the following relation applies [365]:

$$\Delta f/f_Q = (f_Q - f)/f_Q = \rho_F l_F/\rho_Q l_Q = m_F/m_Q \qquad (11.3)$$

where ρ_F and ρ_Q are the thicknesses of the film and the quartz, respectively, and m is the density of the surface. This relationship shows that the film-thickness-measuring quartz is primarily a mass-sensitive sensor. For this reason, all the sensors which are based on this principle are known as *quartz microbalance* (QMB) sensors [366]. Equation (11.3) has been shown to be experimentally valid for layer depositions of $m_F/m_Q < 2\%$. However, in practice thicker coats are often needed. A breakthrough in the theoretical understanding of the sensor characteristics of layer-thickness-measuring quartzes at high thicknesses came when the quartz was treated as a coupled, two-layer resonator [367]. It has been shown [368] that the simple analytical form

$$Z_Q \tan Q + Z_F \tan F = 0 \qquad (11.4)$$

where $Q = f/f_Q$, $F = f/f_F$ and $Z_F = \rho_F c_F$, $Z_Q = \rho_Q c_Q$ must be valid. c is the phase

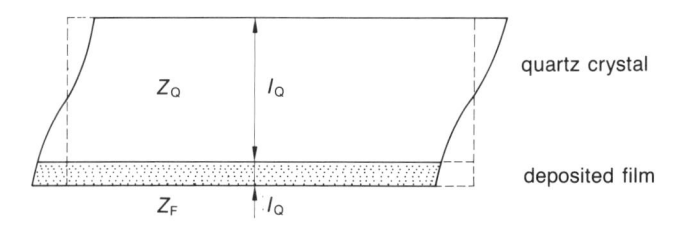

Figure 11.3 One-dimensional model of a quartz microbalance sensor (l_F, l_Q: thickness of film and quartz, respectively; Z_F, Z_Q: sound impedances) [369].

speed of the shear wave in the quartz (subscript Q) or film (F). Using equation (11.2) we obtain

$$m_F/m_Q = \frac{Z_F f_Q}{Z_Q \pi f} \tan^{-1}\left(\frac{Z_Q}{Z_F} \tan \pi \frac{f}{f_Q}\right) \qquad (11.5)$$

This relationship is valid for layer depositions $m_F/m_Q < 70\%$. The sound impedance of the deposited layer must be known [369]. There are many commercial suppliers of this type of sensor.

The mass sensitivity of QMB sensors is being increasingly employed for the detection of chemical and biological substances. The advantages of such sensors are their simple construction, their reliability and, of course, the frequency output they supply. If a chemically or biologically active layer is deposited on the quartz it is possible to produce very sensitive and, in part, very selective new sensors (Figure 11.4). Long-known examples of this type of sensor are gas sensors which use various organic layers as the sensitive layer, and humidity sensors in which water-absorbent layers are applied to the quartz [370–373]. Recently, biological substances such as enzymes or antibodies have been the subject of particular interest [290, 366]. These sensors make use of AT-cut quartzes with frequencies of 9–15 MHz. Sensitivities of 2600 Hz/µg can be achieved using a 15 MHz crystal. If two quartzes are used, one of which is a reference quartz and is not exposed to the medium, then a frequency resolution in the hertz range is attainable. The detection limit is

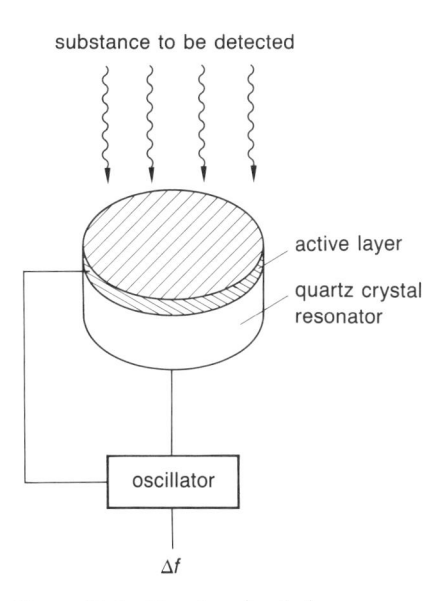

Figure 11.4 Quartz microbalance sensor.

190 Resonance sensors

approximately 10^{-12} g. Monomolecular layers or single 'large' molecules, such as biomolecules, can be detected. This technique can be used in gaseous media without the need for special measures. In liquid media it is necessary to introduce a gas-permeable membrane in front of the quartz, since the oscillation of the quartz is highly damped in liquids with a consequent loss of sensitivity. Berg et al. [375] present a design which uses highly complex electronics and a 25 MHz quartz oscillator and which can be used in liquids directly. If this technique can be made to work in practice, then it will represent a dramatic breakthrough.

The most spectacular application so far is the use of this type of sensor to detect cocaine and morphine [374]. Such devices will make it possible to replace police sniffer dogs.

It has long been known that quartz is a suitable material for temperature sensors [364]. In accordance with Figure 11.2, the HT cut should be preferred since it exhibits a maximum temperature coefficient of approximately 9×10^{-5} K^{-1}. The quartz frequency changes by approximately 0.01% K^{-1}. Although this is a small effect, the excellent stability and precision of the quartz mean that it can be easily evaluated using digital methods. In theory, this type of sensor can be used in a range extending from a few kelvin up to the Curie temperature of quartz (573°C). However, in practice today's encapsulation technology only permits temperatures between $-40°$C and approximately 300°C. It was not until the development of microprocessors that this type of sensor began to play a significant role [364] since, as mentioned above, the precise temperature dependence of the resonant frequency is given by a cubic parabola:

$$f = f_0(1 + \bar{\alpha}T + \bar{\beta}T^2 - \bar{\gamma}T^3) \tag{11.6}$$

where f_0 is the temperature at $T = 0°$C. For the HT cut, $\bar{\alpha} = 9 \times 10^{-5}$ K^{-1},

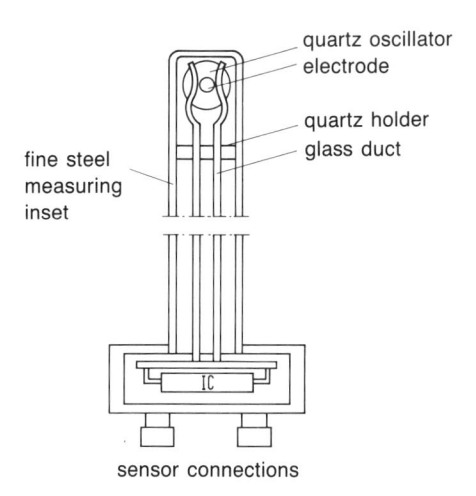

Figure 11.5 Quartz temperature sensor [376].

$\bar{\beta} = 6 \times 10^{-8}$ K^{-1}, $\bar{\gamma} = 3 \times 10^{-11}$ K^{-1}. This means that the non-linearity $\bar{\beta}/\bar{\alpha}$ at high temperature ranges has to be corrected. Nowadays this can be done easily using microprocessors. However, it has been necessary to solve other problems such as the difference in the null point and sensitivity between different samples, the influence of environmental temperatures and self-heating as well as of drift, ageing and hysteresis. As a result, it was not until 1986 that the first temperature sensor based on this method became commercially available. This was System QuaT, manufactured by Heraus [376, 377]. Figure 11.5 illustrates the construction of this sensor. The sensor electronics take the form of a customer-specific circuit (Figure 11.6). The electronics step the oscillator frequency of 16 MHz down to 2 Hz. In order to save power, this signal is transmitted not as a square pulse but in short pulses of less than 1 ms duration at a nominal interval of 0.5 s (Figure 11.7). The variation in this time interval is then temperature-dependent. The changes can be measured digitally. This type of pulse-interval-coded signal offers a number of advantages. It is possible to connect several (for example, up to 16) sensors of this type at a two-core bus line. The pulse intervals can be made to alternate in such a way that the measurement information from all 16 sensors can be transmitted in a single measurement cycle (Figure 11.7). The measuring accuracy of this type of sensor in the above-mentioned temperature range is 0.2 K. This interesting type of sensor shows that the skilful and constructive use of new electronic techniques can return well-known physical effects to the centre of attention.

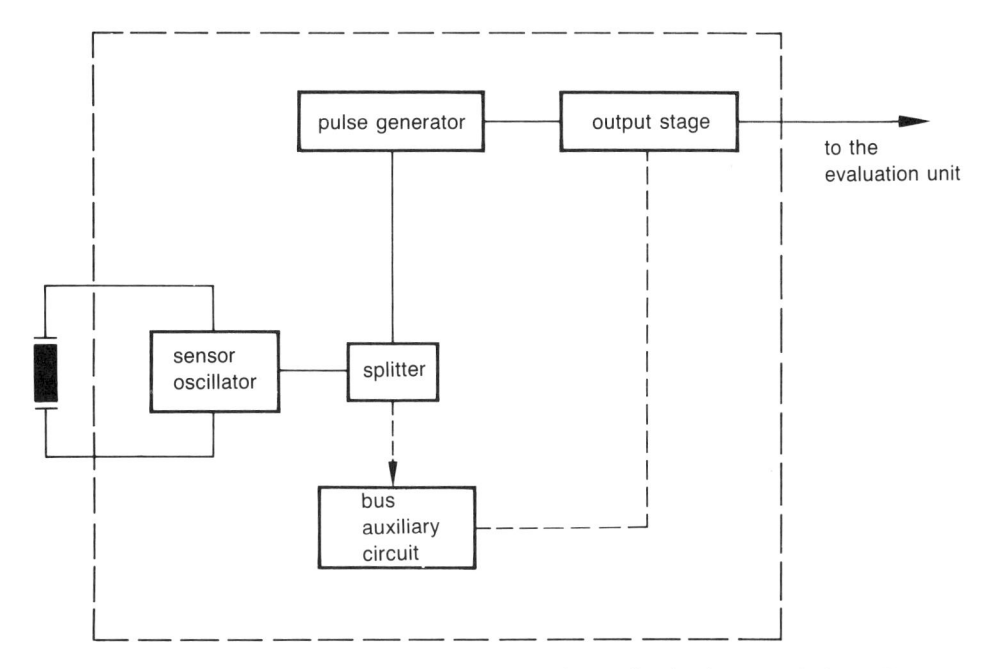

Figure 11.6 Block diagram of sensor electronics realized using an ASIC [376].

192   Resonance sensors

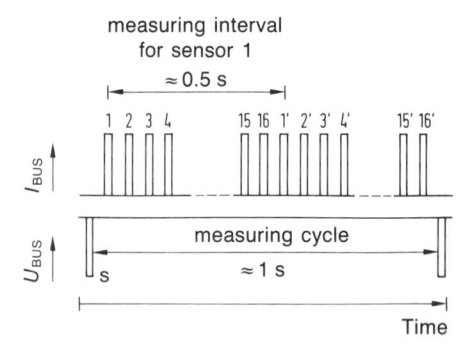

Figure 11.7   Signal sequence in bus operation [376].

Quartz-based bending oscillation resonators are of interest for a certain number of practical applications. Geometrically similar to a two-pronged or three-pronged hybrid transformer, they react to the influence of force with a change in the frequency of the system. They are used in acceleration sensors or in balance systems [364].

11.2 Surface acoustic wave sensors

Surface acoustic waves (SAWs) have long been used for certain electronic components, such as for delay lines, analog and digital filters and resonators. This type of component is well protected from the environment. Surface acoustic waves are mechanical waves at the surface of a piezoelectric solid. The most important SAWs are the *Rayleigh waves*. These possess a longitudinal component which propagates along the plane surface and a vertical shear component which decreases exponentially in the substrate. The speed of propagation of this type of wave is approximately 10^{-5} times the speed of light. These waves are generated by affixing two pairs of comb-shaped, interlocking metal electrodes at the two ends of a piezoelectric substrate using lithographic methods (Figure 11.8). The distance between neighbouring 'teeth' is adjusted to half the wavelength of the surface acoustic wave that is to be produced. An electric signal applied to one of the pairs of electrodes generates an electric field in the piezoelectric substance which causes deformation waves on the surface which propagate in both directions perpendicularly to the teeth of the comb structure. The wave which travels across the substrate plane to the other pair of electrodes can be detected there as an electric signal. The frequency of the SAW can lie between 10 MHz and 1 GHz. The wavelength is between 1 and 100 µm and is also dependent on the properties of the substrate. The wave amplitude typically lies in the angstrom range. The special feature of SAWs is that their propagation is limited to a thin layer along the surface

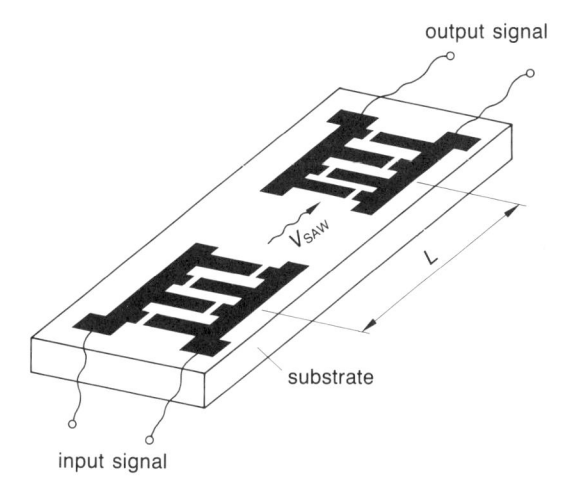

Figure 11.8 SAW element.

of the substrate, the thickness of which is approximately the same as the wavelength of the SAWs. They can be coupled to an adjoining medium which can be gaseous, liquid or solid. At the same time, this coupling exerts a considerable influence on the amplitude and speed of propagation of the SAWs. This means that it is possible to determine the coated mass or the properties of this adjoining layer directly. It is also possible to detect changes in conductivity or dielectric constant.

Quartz, lithium niobate, piezoceramics or thin piezoelectric layers such as ZnO can be used as the piezoelectric substrate material. These thin layers are located on a non-piezoelectric base such as glass, silicon or a ceramic [378]. What is remarkable about SAW-based components is that their performance increases considerably as the frequency rises.

Alongside Rayleigh waves, other waves such as *Lamb waves*, *Love waves* and *Bluestein–Gulyaev* waves are of interest as SAWs [369]. For example, Lamb waves are important when propagation in thin membranes is required.

Surface acoustic wave sensors make use of the influence on the propagation parameters of the SAWs of environmental quantities (T, p), the applied layers, fields or changes in the distance between the electrodes caused by external stresses or deformations. Although these devices are more expensive than sensors manufactured using other technologies, for example silicon-based temperature or pressure sensors, they are attracting increasing attention on account of the exceptional properties which they possess. One of these is the wide dynamic operating range, coupled with excellent linearity. This makes them attractive for precision measurements. Another advantage lies in their frequency-analog output signal. They are used as delay lines in oscillator circuits or as resonators. The frequency is directly proportional to the speed of the SAWs. Thus a SAW delay line

194 Resonance sensors

in an oscillator circuit (Figure 11.9) oscillates at a stable frequency f if the following condition is fulfilled:

$$f = (2n\pi - \varphi_{el}) v_{SAW}/L \tag{11.7}$$

where φ_{el} is the electrical phase shift of the amplifier circuit, and n is a constant whose value is limited by suitable design of the electrodes and the bandwidth of the HF amplifier. As a result, f is dependent on v_{SAW} and L. This makes it possible to develop sensors for the detection of acoustic, biological, mechanical and thermal quantities. A number of examples are presented below [379, 380].

In temperature measurements, the temperature dependence of the speed of propagation of the SAW is used. The important properties here are a high degree of temperature dependence and as close an approximation to linearity as possible. The anisotropy of the usual substrate materials means that this can be influenced by the choice of the crystal cut and the direction of propagation of the wave. However, temperature does not only influence the speed of propagation of the SAW but also the physical dimensions. $LiNbO_3$ possesses a relatively large temperature coefficient which results in a change in delay time of 94 ppm K^{-1}. Handen et al. [382] present a report of a temperature sensor based on a JCL quartz in an oscillator circuit with a resolution of the order of thousandths of a degree, high linearity and low hysteresis. It can operate within the range $-100°C$ to $+200°C$. Frequently temperature dependence is an undesired effect. In this case two delay lines are used to eliminate the temperature drift.

Pressure sensors, for example for gas measurement, use quartz membranes (Y cut) with a thickness of 100–200 μm on the front of which SAW resonators are attached. While the pressure-sensitive resonator configuration is located in the centre of the membrane, the configuration at the edge is used for temperature compensation purposes (Figure 11.10). A resonant oscillator, which achieves better results than a delay oscillator, is used to evaluate the difference in frequency. A pressure sensitivity of 360 Hz/kPa at a resonant frequency of 130 MHz can be achieved [383]. It is also possible to configure force and acceleration sensors [385]. In the case of force sensors for example, delay lines are attached to the top and bottom of a Y-quartz substrate. While one end of the substrate is fixed, a force is

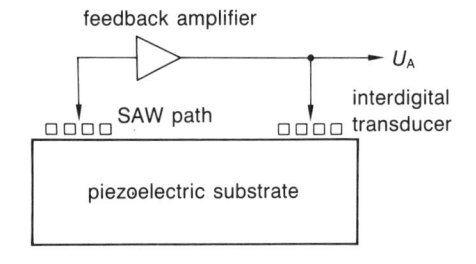

Figure 11.9 SAW delay line oscillator [378].

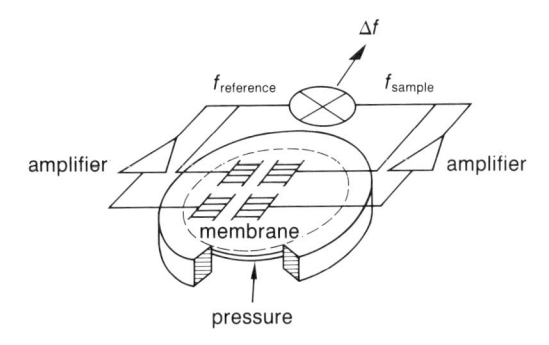

Figure 11.10 SAW pressure sensor [379].

applied at the other. This causes the bar-like substrate to bend. The difference in frequency between the two delay lines provides a measurement of the force. A sensitivity of 1.8 kHz/N can be achieved [384].

When an electric field is applied perpendicular to the surface of a piezoelectric substrate along which a SAW is propagating, the speed of propagation of the SAW changes. This type of sensor can be used for high-tension measurements. Its advantages consist in its extremely high input impedance and its good dielectric insulation.

Surface acoustic wave sensors are also used as displacement sensors [387], flow sensors [388] or sensors for the characterization of films, for example for film deposition [389] or changes in films [390]. Since the late 1970s much work has concentrated on the use of SAW sensors as chemical or biological sensors [379]. Humidity and condensation point sensors also belong to this category. The principle underlying most of these sensors is that a chemically sensitive layer which has been applied to a piezoelectric substrate absorbs the species that is of interest. This leads to a change of mass which can, experimentally, be determined as a change in frequency. Figure 11.11 illustrates an H_2 sensor which uses a ZnO thin film as the piezoelectric material and Pd as the sensitive substance [378]. It is essential to note that the carrier material for this sensor is silicon and that as a consequence a monolithic, integrated sensor can be constructed.

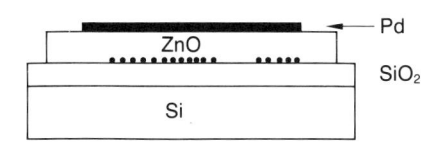

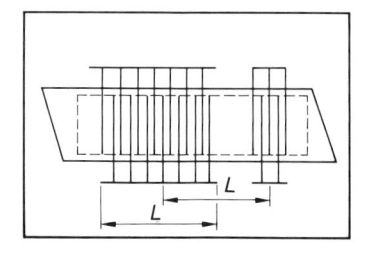

Figure 11.11 H_2 SAW sensor [378].

Using ST quartz SAW sensors, which operate at 158 MHz and possess an active surface of 0.08 cm^2, sensitivities of 400 Hz/ng have been achieved [379]. The same materials that are used in QMB sensors can of course also be applied to SAW sensors. However, SAW sensors are more sensitive than the bulk oscillators. As for the question of achieving a selective response, chemical and biological SAW sensors are subject to the same problems that have been listed elsewhere.

It is to be expected that the use of these sensors will be extended to further areas of application in the future. It is thus possible to produce magnetic field sensors by depositing magnetostrictive material on the surface of an SAW sensor. The use of pulse–echo techniques extends the range of information that can be provided. The use of the many other types of acoustic wave that occur in solids will also prove to be of great interest. This will make it possible to produce sensors with improved sensitivity. An example is given by the latest research into a multisensor based on a Lamb wave oscillator [391]. At the heart of this sensor is an ultrasound delay line which consists of a thin membrane made of a ZnO/Al/Si$_x$N$_y$ composite (Figure 11.12). The membrane is etched from a silicon substrate and its thickness is much smaller than the wavelength of the ultrasonic wave that propagates in it, amounting to some 3 µm at a length of approximately 1 mm. Lamb waves can form in this type of membrane. Sensors for force, acceleration, pressure and for fluid properties such as viscosity and density can be produced on this basis. They are also attractive as gravimetric sensors for the detection of chemical or biological species and possess a number of advantages compared to QMB or even SAW sensors. Table 11.1 compares the three types [392]. The most important conclusions are as follows:

- The sensitivity of QMB and SAW sensors increases with frequency (decreasing λ); in Lamb wave sensors (LWSs) sensitivity increases with decreasing thickness d (and consequently with falling frequency f, as $f_{LW} \sim d^{3/2}$).

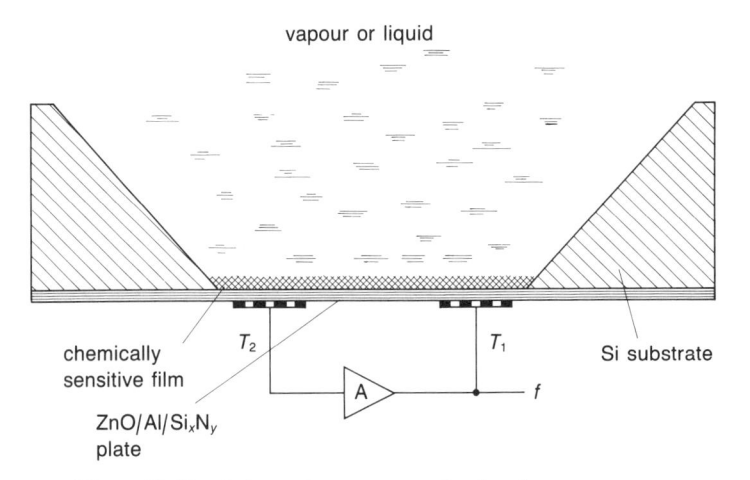

Figure 11.12 Schematic structure of a Lamb wave sensor.

Density, filling level and throughflow   197

Table 11.1 Comparison of QMB, SAW and Lamb wave sensors

Sensor	Theoretical sensitivity S	Principle	Frequency (MHz)	S_{ber} (cm^2/g)
Quartz microbalance (Bulk)	$-2/(\rho\lambda)$	AT quartz resonator	6	-14
SAW	$-K(\sigma)/(\rho\cdot\lambda)$	ST SAW delay time	112	-151
Lamb waves (LW)	$-1/(2\rho d)$	ZnO/Si_xN_y	4.7	-450

$K(\sigma) = 0.8...2.2$; $S = \lim_{\Delta m \to 0} 1/f(\Delta f/\Delta m)$; ρ – Density, λ – Wavelength, d – Thickness of membrane

- The ratio of the sensitivities of QMB, SAW and LWS is 1:10:40 when the frequencies of QMB and LW are equal and the frequency of the SAW is 20 times higher.

At first glance it would seem that the LWSs should be preferred to the others. They possess the highest level of sensitivity, operate at a 'comfortable' frequency and are equally well suited as, for example, chemical sensors for gases and liquids. However, these advantages are countered by the fact that they are complicated to manufacture. Surface acoustic wave sensors can be constructed from commercially available components. However, they suffer from the fact that they operate at high frequencies. Finally, the disadvantage of QMB sensors is that they are less sensitive. On the other hand, they possess a large sensing surface and are cheap. The future will show which advantages prove the most important.

11.3 Resonance sensors for density, filling level and throughflow

One extreme approach to the presentation of sensors would be only to describe miniaturized sensors. Another would be to try to describe the entirety of measurement principles and procedures. The following examples could form part of this second extreme. They should nevertheless be mentioned at this point since they complete our discussion of the principles of resonant sensors. We are concerned here with methods employed in industry to produce a frequency-analog output signal [364]. A resonant sensor for measuring the density of liquids is presented in Figure 11.13. A tube, through which the liquid flows, is 'divided' into two parts within a limited region. This region is mechanically separated from the rest of the tubing system by a flexible connection. This eliminates the influence of vibrations at the fixing points. An external influence causes the two parts to oscillate in such a way

198 Resonance sensors

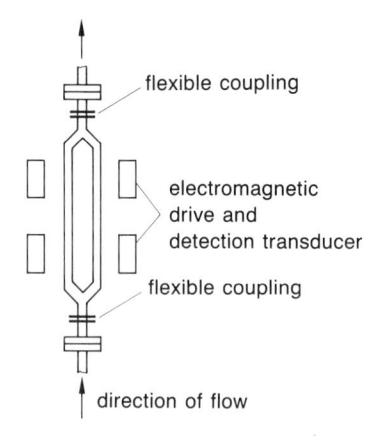

Figure 11.13 Resonant densitometer.

that they are always moving in opposite directions. The resonant frequency of the excited system now depends on the density of the liquid:

$$f = f_0/(1 + \rho/\rho_0) \tag{11.8}$$

where f_0 is the resonant frequency without liquid in the tube and ρ_0 is a constant which is dependent on the tube geometry. The change in frequency f is continuously measured using a second pair of transducers which record the size of the oscillations. This resonant sensor operates with an accuracy of 0.1% and can be used for a wide variety of liquids. Knowledge of the density makes it possible to deduce the concentration or composition of the liquid.

Figure 11.14 shows a resonant filling level sensor. A long tube is brought to a state of bending oscillation by the action of an internal piezoelectric transducer. If this tube is immersed in a liquid, then the resonant frequency changes as a function of the depth of immersion. Typical configurations of this type of sensor function at frequencies in the kilohertz range. Such devices are used in aircraft and in fuel transport trucks.

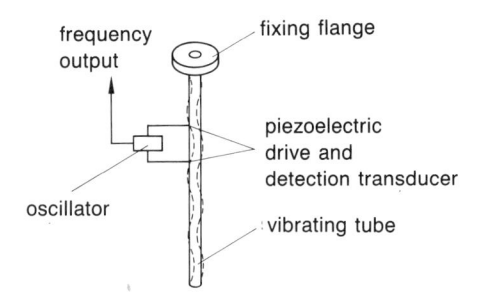

Figure 11.14 Filling level sensor based on the resonance principle.

A new method of measuring throughflow which must be numbered among the resonance techniques is the Coriolis flowmeter, shown in Figure 11.15(a). It consists of a U-shaped tube with a free front end which is caused to oscillate in the way depicted by a (for example, electromagnetic) transducer. If no mass is flowing ($q = 0$) then the oscillation mode remains constant. However, if a medium is flowing with velocity **v**, then Coriolis forces

$$\mathbf{F}_C = 2m(\omega \times \mathbf{v}) \tag{11.9}$$

occur (Figure 11.15(b)). Depending on the direction of flow, these forces act in different directions at staggered points along the tube and generate a torque M:

$$M = 2F_C r = K\alpha \tag{11.10}$$

ω in equation (11.9) is the angular frequency of the excitation oscillation, and in equation (11.10) $2r$ is the length of the front of the U-shaped tube and K is the angular momentum. Depending on its direction, the Coriolis force F_C periodically 'rotates' the tube through the angle α (Figure 11.15(b)). This can be measured using sensitive transducers. α provides a direct measure of the mass flow:

$$q_m = (K/4)r\omega\alpha \tag{11.11}$$

This direct recording of the mass flow eliminates the otherwise frequent need to convert the volume into mass. Pressure and viscosity are of no significance and changes in density, gas, impurities and pulsations in flow only intervene outside

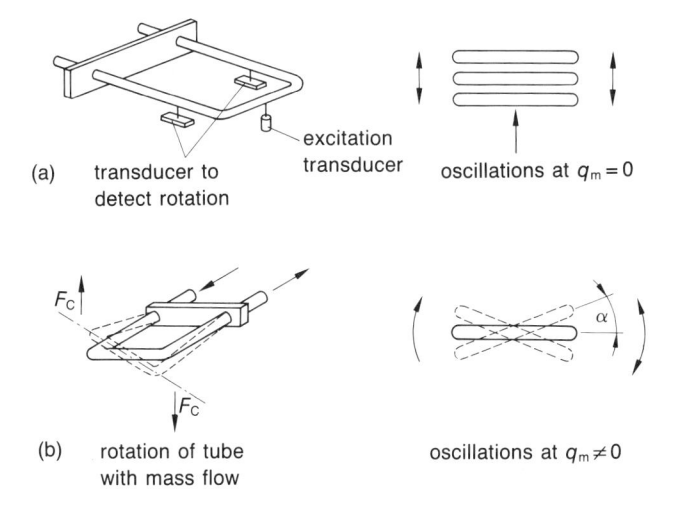

Figure 11.15 (a) Coriolis flowmeter; (b) and its behaviour in a flowing medium.

certain limits. The procedure is particularly well suited for the dosage of very small and medium-sized amounts at a measuring accuracy of 0.5%. In practice, many different configurations of this U-type sensor are in use. However obvious the technique may be, its realization still requires considerable technical effort [393]. The frequency of oscillation can amount to 80 Hz at amplitudes of 1 mm. At low masses the Coriolis deformation amounts to approximately 10 µm. This means that the principle cannot be used over large nominal distances.

12 Prospects for the future

In the near future sensor technology will be able to make even greater use of the technological potential of microelectronics. This means that there will be more and more sensors based on microelectronic technologies such as semiconductor, thin- and thick-film technologies. Estimates predict that 40–60% of all sensors in the 1990s will be semiconductor-based, making use of silicon in particular. This development also includes the integration of evaluation and processing electronics. Monolithic integration, for example of amplifiers, multiplexers and voltage–frequency converters is a possibility. This makes available a frequency-analog signal which is easily processed. However, it would be better still if the sensor itself could provide this information directly. A second step would thus be to integrate a microcomputer. However, for cost reasons this is only likely to be realized in special cases. This variant would have the advantage of providing a bus-compatible signal, and the first examples of this type of design have emerged. However attractive the principle of monolithic integration might appear, it still has disadvantages. The sensor and the integrated electronics can affect each other in undesirable ways, for example through the dissipation of heat from the sensor to the electronics. Encapsulation represents a considerable problem. The technology of the sensor and the circuitry can differ significantly, with the sensor requiring the best possible contact with the environment from which the electronics has to be shielded. These problems are being examined intensively. The problems are slightly less pressing when hybrid techniques are used or the electronics is housed separately. In the latter case, SMD technology and the use of ASICs are becoming increasingly important. Single-chip computers are being used to correct drift, offset, non-linearity, etc., and to improve the quality of the signal. In such cases we talk of *integrated* sensors.

The future will also see micromechanical sensors playing an important role. Ideally these would unite the sensor, processing electronics and actuator on a single chip. It is to be expected that optical fibers will become cheaper as the technologies involved in their production become less costly. With advances in the field of integrated optics and the coupling of other technologies, such as micromechanics, it can be expected that many interesting new fiber optic applications will emerge. However, the use of optical fibers in sensor applications is only just beginning.

The field of chemical sensors is benefiting considerably from microelectronics technology (coating, contacting) on the one hand, and from developments in the materials field (substrate, sensitive layer, etc.) on the other. Many new and further developments can be expected.

Development in the sensor field is not fuelled simply by technologies and techniques but also by new materials such as polymers, ceramics, special semiconductors, metallic glasses and, possibly, superconductors.

To a large extent sensor development is also stimulated by the variety of applications for which sensors are required. The range is growing constantly and extends into every sector of daily life. The focus of interest is not separate sensors but, increasingly, sensor systems. Such systems can be composed of sensors of the same (redundant system) or different (diversified system) types. In conjunction with the required computer processing, this interconnection of sensors is opening up a whole new range of possibilities. Pattern recognition is an excellent example of this.

Given the current state of computer development, the digital sensor would be the ideal product. However, in the foreseeable future it is an ideal which, without the use of intermediary electronics, will only be realized in exceptional cases. It is possible that application-specific integrated sensors (ASISs) will play an important role in the future. These devices accommodate a large number of sensors of the same or different types on a single chip. The configuration can be adapted to specific customer requirements. A chip developed with this objective in mind and containing 64 sensors was first developed at Berkeley University in 1986 [395].

References

1. Medlock, R. S.: *Measurement and Control* 20 (1987), 6.
2. Schmidt, B. and Schubert, D.: *Siliciumsensoren*. Akademie-Verlag, Berlin 1986.
3. Juckenack, D.: *Handbuch der Sensortechnik*. Verlag moderne Industrie, Landsberg/Lech 1989.
4. Heywang, W.: *Sensorik*. Springer-Verlag, Berlin, 1984, 2nd edn 1986.
5. Reichl, H. et al.: *Halbleitersensoren*. expert-Verlag, Ehningen 1989.
6. Schanz, G.W.: *Sensoren*, Hüthig-Verlag, Heidelberg 1986.
7. Wiegleb, G.: *Sensortechnik*, Franzis-Verlag, Munich 1986.
8. Bonfig, K. W., Bartz, W. and Wolff, J.: *Sensoren, Messaufnehmer*. expert-Verlag, Ehningen 1988.
9. Ahlers, H. and Waldmann, J.: *Mikroelektronische Sensoren*. Verlag Technik, Berlin 1990.
10. Hart, K.: *Feingerätetechnik* 31 (1982) 10, 470.
11. Günzel, K.: *Technisches Messen* 50 (1983) 10, 355.
12. Tränkler, H. R.: *Technisches Messen* 49 (1982) 10, 343.
13. Tränkler, H. R.: *messen prüfen automatisieren* June (1986), 332.
14. Richter, W.: *Grundlagen der elektrischen Messtechnik*. Verlag Technik, Berlin 1985.
15. Richter, W.: *Feingerätetechnik* 34 (1985) 1, 2.
16. Jäger, G. and Grünwald, R.: *Feingerätetechnik* 36 (1987) 7, 294.
17. Tschulena, G. and Selders, M.: *Technisches Messen* 50 (1983) 4, 127.
18. Recherche Sensordatenbank, *Sensor-Special 86/87*. VDI-Verlag.
19. Middelhoek, S. and Noorlag, D.J.W.: *NTG Fachbericht* 79 (1982) 35.
20. Germer, W.: *VDI-Bericht* 509 (1984), 209.
21. Obermeier, E. and Reichl, H.: *NTG-Fachbericht* 79 (1982), 49.
22. Weissel, R.: *NTG-Fachbericht* 79 (1982), 56.
23. Wieser, E.: *ZfK-Bericht* KFHI.83, Rossendorf (1983).
24. Binder, J. et al.: *Sensors and Actuators* 4 (1983), 527.
25. Heuberger, A.: *Physikalische Blätter* 43 (1987) 1, 14.
26. Benecke, W.: *hard and soft. Fachbeilage Mikroperipherik* February (1986), 33.
27. Angell, J. B., Terry, S. C. and Barth, P. W.: *Silicon Micromechanical Devices*, Scientific America 4 (1983).
28. Meek, R. L.: *J. Electrochem. Soc.* 118 (1971) 437.
29. Sackson, T. N., Tischler, M. A. and Wise, K. D.: *IEEE Electr. Dev. Lett.* 2 (1981) 437.
30. Allen, R.: *Electronic Design* 34 (1986) 26, 71.
31. Paul, R.: *Halbleiterdioden*. Verlag Technik, Berlin 1976.

32. Müller, R. and Heywang, H.: *Grundlagen der Halbleiterelektronik.* Springer-Verlag, Berlin 1979.
33. O'Neil, P. and Derrington, C.: *Elektronik* 29 (1980) 11, 81.
34. McNamara, A. G.: *Rev. of Sc. Instr.* 33 (1962) 3, 330.
35. Abu-Zeid, M. M. and Sasburg, A. G.: *Rev. of Sc. Instr.* 57 (1986) 10, 2609.
36. Hwang, H. J. and Reichl, H.: *FhG-Berichte* 3 (1981) 11.
37. Zickler, A., Elbel, Th. and Müller, J.: *Proc. SENSOR 88*, Part A, Nuremberg 1988, p. 83.
38. Van Herwaarden, A. W.: *Thermal Vacuum Sensors Based on Integrated Si-Thermopiles*, Dissertation, Delft (Nd) 1987.
39. Kodato, S. et al.: *J. Non-crystalline Solids* (1983), 1207.
40. Takahshi, K.: *Proc. TRANSDUCER 1987*, Tokyo, p. 235.
41. Mason, W. P.: *Crystal Physics of Interaction Processes.* Academic Press, New York 1966.
42. Zerbst, M.: *Piezoeffekte in Halbleitern. Festkörperprobleme II.* Vieweg, Braunschweig 1963.
43. Kanaa, Y.: *IEEE Trans. Electr. Dev.* ED-29 (1982) 64.
44. Lenk, A.: *Elektromechanische Systeme* Vol. 3, Verlag Technik, Berlin 1975.
45. Bradshaw, A.T.: *rtp* (1983) 12, 531.
46. Conrads, W.: *EEE, Elektronik-Appl.* October (1985), 74.
47. Hellwig, R.: *mpa* May (1986), 273.
48. Reichl, H. et al.: *Sensors and Actuators* 4 (1983), 247.
49. Sugiyama, S., Takigawa, M. and Igarashi, I.: *Sensors and Actuators* 4 (1983), 113.
50. Neumeister, J., Schuster, G. and von Munch, W.: *Sensors and Actuators* 7 (1985), 167.
51. Colman, D., Bate, R. T. and Mize, J. P.: *J. Appl. Phys.* 39 (1968), 1923.
52. Mikoschiba, H.: *Solid-State Electr.* 24 (1981), 221.
53. Canali, C. et al.: *J. Phys.* D 12 (1979), 1973.
54. Dorda, G.: *J. Appl. Phys.* 42 (1971), 2053.
55. Schörner, R.: *VDI-Berichte* 677 (1988), 93.
56. Lian, W. and Middelhoek, S.: *Sensors and Actuators* 9 (1986), 259.
57. French, P. J. and Lian, W.: *Proc. EUROSENSORS I*, Cambridge (UK) 1987, p. 213.
58. Germer, W.: *VDI-Berichte* 509 (1984), 209.
59. Shoji, S., Nisase, T. and Esashi, M.: *Proc. TRANSDUCER 1987*, Tokyo, p. 305.
60. French, P. J., Lian, W. and Middelhoek, S.: *IEEE Proc.* 135, Part D (1988) 5, 359.
61. Miyoshi, S. et al.: *Proc. TRANSDUCER 1987*, Tokyo, p. 309.
62. Schmidt, B. and Schubert, D.: *Proc. 5th Intern. School on Physical Problems in Microelectronics*, Varna 1987.
63. Breitner, M. and Tomasi, G.: *Siemens Forsch.- u. Entwicklungsbericht* 10 (1981) 2, 65.
64. Voorhuyzen, J. A. and Bergveld, P.: *IEEE Trans. Electron. Dev.* ED-32 (1985) 7, 1185.
65. Sprenkels, A. J., Voorhuyzen, J. A. and Bergveld, P.: *Sensors and Actuators* 9 (1983), 59.
66. Voorhuyzen, J. A. and Bergveld, P.: *Proc. TRANSDUCER 1987*, Tokyo, p. 328.
67. Kawamura, Y. et al.: *Proc. TRANSDUCER 1987*, Tokyo, p. 283.
68. Van Herwaarden, A. W. and Vellekoop, M. J.: *Proc. EUROSENSOR II*, Enschede, Netherlands (1988) p. 155.
69. Keller, H. W. and Anagnostopoulus, K.: *Proc.TRANSDUCER 1987*, Tokyo, p. 316.

70. Bleicher, M.: *Halbleiter-Optoelektronik*. Verlag Technik, Berlin 1986.
71. Adler, D.: *Naturwissenschaften* 69 (1982), 574.
72. Yoshida, A., Setsune, K. and Hirco, T.: *Proc. TRANSDUCER 1987*, Tokyo, p. 249.
73. Koike, F., Okamoto, H. and Hamakawa, Y.: *Proc. TRANSDUCER 1987*, Tokyo, p. 223.
74. Wolffenbüttel, R. F.: *Proc. TRANSDUCER 1987*, Tokyo, p. 219.
75. Yang, D., Ambo, K. S. and Holm-Kennedy, J. W.: *Sensors and Actuators* 14 (1988), 69.
76. Scharff, W. et al.: *Wissens. u. Fortschritt* 38 (1988) 10, 252.
77. Brown, H. T.: *GEC J. Sc. Techn.* 43 (1977), 125.
78. Tsukada, T. et al.: *IE DM* 81-479 (1981).
79. Foner, S.: *IEEE Trans. Magn.* MAG-17 (1981), 3358.
80. Baltes, H. P. and Popovic, R. S.: *Proc. IEEE* 74 (1986) 8, 1107.
81. Stafeev, V. I. and Karakushan, E. I.: *Magnetodioden*, Nauka, Moscow 1975.
82. Takamiya, S. and Fujikawa, K.: *IEEE Trans. Electr. Dev.* ED-19 (1972), 1085.
83. Huang, R. M., Yeh, F. S. and Huang, R.: *IEEE Trans. Electr. Dev.* ED-31 (1984), 1001.
84. Popovic, R. S.: *IEEE Electr. Dev. Letters* EDL-5 (1984), 357.
85. Gallagher, R. C. and Corah, W. S.: *Solid-State Electr.* 9 (1965), 168.
86. Yagi, A. and Sato, S.: *Jpn. J. Appl. Phys.* 15 (1976), 655.
87. Popovic, R.S.: *Solid-State-Electr.* 28 (1985), 711.
88. Hirata, M. and Suzuki, S.: *Proc. 1st Sensor Symp., Inst. Elec. Eng. Japan* (1982), p. 37.
89. Suhl, H. and Shockley, W.: *Phys. Rev.* 75 (1949), 1617.
90. Lotes, O. S. et al.: *IEEE Trans. Electr. Dev.* ED-27 (1980), 2156.
91. Vinal, A. W. and Masnari, N. A.: *IEEE Electr. Dev. Lett.* EDL-3 (1982), 203.
92. Zieren, V., Kordic, S. and Middelhoek, S.: *IEEE Electr. Dev. Lett.* EDL-3 (1982), 394.
93. Vinal, A. W. and Masnari, N. A.: *IEEE Electr. Dev. Lett.* EDL-3 (1982), 396.
94. Davies, L. W. and Wells, M. S.: *Proc. IREE Australia (1971)*, 235.
95. Kordic, S.: *Electr. Dev. Lett.* EDL-7 (1986), 196.
96. Persky, G. and Bartelink, D. J.: *Bell. Syst. Techn. J.* 53 (1974), 467.
97. Goicolea, J. I. et al.: *IEEE Trans. Electr. Dev.* ED 28 (1981), 1252.
98. Gilbert, B.: *Electr. Lett.* 12 (1976), 608.
99. Kirby, S.: *Sensors and Actuators* 4 (1983), 25.
100. Kirby, S.: *Sensors and Actuators* 3 (1983), 33.
101. *mpa* 7/8 (1986), 394.
102. v. Borcke, U. and Hini, P.: *Siemens-Z.* 47 (1973), 491.
103. Sogiyama, Y., Taguchi, T. and Tacano, M.: *Proc. TRANSDUCER 1987*, Tokyo, p. 547.
104. Heuberger, A.: *Mikromechanik*. Springer-Verlag, Berlin, 1989.
105. Benecke, W.: *Proc. COMPEURO 1989*, Hamburg, p. 3.
106. Wise, K. D.: *Micromachining and Micropackaging of Transducers*, Elsevier, Amsterdam 1985.
107. Angell, J. B. et al.: *Silicon Micromechanical Devices*, Sc. America 4 (1983).
108. Geller, M.: *SENSOR report* 4 (1988), 11.
109. Tsugai, M. and Bessho, M.: *Proc. 10th Vehicle Automation Symp., JAACE* (1987), p. 71.
110. Rudolf, F. et al.: *Proc. TRANSDUCER 1987*, Tokyo, p. 395.

111. Lyman, J.: *Electronics* 60 (1987) 20, p. 85.
112. Howe, R. T. and Müller, R. S.: *IEEE Trans. Electr. Dev.* ED-33 (1986) 4, 499.
113. Poros, J. M.: *ISA 21st Int. Instr. Symp. Philadelphia*, May 1985.
114. Allen, R.: *Electron. Design* 34 (1986) 26, 71.
115. Hegner, F.: *Techn. Messen* 52 (1985) 10, 379.
116. Dössel, O. and Graeger, V.: *Elektrotechnik* 67 (1985) 23/24, 23.
117. Germer, W.: *Sensors and Actuators* 7 (1985), 135.
118. Tencey, A.: *I & CS* August (1987), 21.
119. Godefray, J. C. et al.: *J. Vac. Sci. Technol.* A 5 (1987) 5, 2917.
120. Loreit, U. et al.: *radio fernsehen elektronik* 43 (1985) 5, 316.
121. Dibbern, V.: *industrie electric-electronic* 28 (1983) 6, 36.
122. *EEE-Elektronik-Applikation* 16 (1986) 9, 44.
123. Anderson, J. C.: *J. Vac. Sci. Techn.* A 4 (1986), 610.
124. Just, H. J. and Sichting, T.: *Feingerätetechnik* 34 (1985) 7, 310.
125. Janoska, I. and Haskara, M. R.: *Sensors and Actuators* 8 (1985), 3.
126. Storch, W. et al.: *radio fernsehen elektronik* 37 (1988) 2, 71.
127. Dell'Acqua, R. et al.: *Proc. 3rd European Hybrid Microelectronics Conference 1981*, Avignon, p. 121.
128. Graeger, V. and Liehr, M.: *atp* 10 (1985), 476.
129. Iwanga, S. and Ikegami, A.: *Proc. 31st Electronic Components Conf.* 1981, p. 58.
130. Channon, N. D.: *J. Soc. Environmental Eng.* 18 (1979), 23.
131. Leppävuori, S.: *Electrocomponent Sc. and Technology* 6 (1980), 193.
132. Giallorenzi, T.G.: *IEEE Trans.on Microwave Theory and Techniques* MTT-30 (1982) 4, 472.
133. Kersten, R. T.: *Physik in unserer Zeit* 15 (1984) 5, 139.
134. Marcuse, D.: *Principles of Opt. Fibre Measurements*, Academic Press, New York 1981.
135. Dors, R.: *Umschau* (1986) 8, 416.
136. Culshaw, B.: *Optical Fibre Sensing and Signal Processing*, Peter Peregrinus, London 1984.
137. Holst, A. et al.: *Feingerätetechnik* 35 (1986) 11, 495.
138. Medlock, R. S.: *Measurement and Control* 19 (1986) 1, 3.
139. Hampartsoumian, E. and Williams, A.J.: *Inst. of Energy* Dec. (1985), 159.
140. Johnson, M.: *Opt. Engineering* 24 (1985) 6, 961.
141. Ulrich, R.: *Techn. Messen* 53 (1986) 2, 313.
142. Ulrich, R.: *Techn. Messen* 51 (1984) 6, 205.
143. Grimm, E. and Nowak, W.: *Lichtwellenleitertechnik*. Verlag Technik, Berlin 1988.
144. Hochberg, R. C.: *IEEE Trans. Instr. Measurement* IM-35 (1986) 4, 447.
145. Jones, B. E.: *J. Phys. E: Sc. Instr.* 18 (1985), 770.
146. Martin, M. J. et al.: *Med. and Biol. Engineering and Computing* 25 (1987), 597.
147. Moretti, M.: *Laser Focus/Electro-Optics* May (1987), 118.
148. Scheggi, A. M.: *J. Opt. Sensors* 2 (1987) 1, 3.
149. *Talantia* 35 (1988) 2, 1.
150. Smith, A. M.: *Electronics and Power* Nov./Dec. (1986), 811.
151. Murray, R. T. et al.: *J. Opt. Sensors* 1 (1986) 4, 317.
152. Krohn, D. A.: *Proc. ISA Int. Conf. and Exhibition*, 1982, Philadelphia.
153. Spillmann, W. B. and McMahon, D. H.: *Appl. Phys. Lett.* 37 (1980), 145.
154. Ovren, C., Adolfsson, M. and Hök, B.: *Int. Conf. on Opt. Techn. in Process Control*, 1983, The Hague, p. 67.

155. Wickersheim, K. A. and Alves, R. B.: *Ind. Research/Development* (1979), 82.
156. Grattam, K. T. V. and Palmer, A. W.: *Rev. Sc. Instr.* 56 (1985) 9, 1784.
157. Accufiber Corporation, *Technical Literature*. Vancouver, Washington (1983).
158. Christiansen, C.: *Ann. Phys. Chem.* 23 (1984), 298.
159. Brenci, M. *et al.*: *OFS Conf. Proc. 1984*, Stuttgart, p. 155.
160. Jonsson, L. and Hök, B.: *OFS Conf. Proc. 1984*, Stuttgart, pp. 3, 191.
161. Fritsch, K. and Beheim, G.: *Opt. Letters* 11 (1986), 1.
162. Fehrenbach, G. W.: *SENSOR '88*, Nuremberg p. 49.
163. Culshaw, B.: *The Radio and Electronic Engineer* 52 (1982) 6, 283.
164. Jackson, D. A.: *J. Phys. E: Sc. Instr.* 18 (1985), 981.
165. Culshaw, B.: *J. Opt. Sensors* 1 (1986) 3, 237.
166. Voges, E.: *hard and soft. Fachbeilage Mikroperipherik*, V, June 1987.
167. Eberhardt, F. J. and Andrews, F. A.: *J. Acoust. Soc. Am.* 48 (1970), 603.
168. Giallorenzi, T.G. *et al.*: *J. Quantum Electr.* QE-18 (1982), 626.
169. Kersey, A. D., Jackson, D. A. and Corke, M.: *Electron. Letters* 18 (1982), 559.
170. Corke, M. *et al.*: *Electron. Letters* 19 (1983), 471.
171. Maklad, M. S.: *ISA Trans.* 27 (1988) 1, 25.
172. Kersey, A. D., Jackson, D. A. and Corke, M.: *J. Lightwave Tec.* LT-3 (1985), 836.
173. Memmelstein, M. D.: *Appl. Optics* 22 (1983), 1006.
174. El-Sherif, M. A., Shankar, P. M. and Herczfeld, P. R.: *Proc. TRANSDUCER 1987*, Tokyo, p. 200.
175. Auch, W.: *Techn. Messen* 52 (1985) 5, 199.
176. Stendle, W.: *hard and soft. Fachbeilage Mikroperipherik*, XIV, June 1987.
177. Giallorenzi, T. G. and Bucaro, J.A.: *IEEE Spectrum* September (1986), 44.
178. Burns, W. K. *et al.*: *Opt. Lett.* 9 (1984), 520.
179. Bergh, R. A., Lefevre, H. C. and Shaw, H. J.: *Opt. Lett.* 6 (1981), 198.
180. Eckhoff, W.: *Opt. Lett.* 6 (1981), 204.
181. Gambling, W. A.: *J. Phys. E: Sc. Instr.* 20 (1987), 1091.
182. Varnham, M. P. *et al.*: *Electr. Lett.* 19 (1983), 699.
183. Smith, A. M.: *Appl. Optics* 17 (1978), 52.
184. Buttler, M. A. and Venturini, E.L.: *Appl. Optics* 26 (1987) 9, 1581.
185. Rogers, A. J.: *J. Phys. D: Appl. Phys.* 19 (1986), 2237.
186. Giles, I. P.: *Phys. Technol.* 18 (1987), 153.
187. Dakin, J. P., Wade, C. A. and Withers, P. B.: *Proc. SPIE-522 (Fibre Optics '85)*, London, p. 226.
188. Kist, R.: *Techn. Messen* 54 (1987) 7/8, 304.
189. Theocharous, E.: *Proc. 1st Int. Conf. Optical-Fibre Sensors 1983*, London. IEE Conf. Publ. No. 221, p. 10.
190. Hartog, A. H.: *J. Lightwave Technol.* LT-1 (1983), 498.
191. Dakin, J. P.: *J. Opt. Sensors* 1 (1986) 2, 101.
192. Farries, M. C. *et al.*: *Electron. Lett.* 22 (1986), 418.
193. Dakin, J. P.: *UK Patent Appl.* 2140554A.
194. Rogers, A. J.: *Appl. Opt.* 20 (1981), 1060.
195. Brown, R. G. W.: *J. Phys. E: Sci. Instr.* 20 (1987), 1312.
196. Grattam, K. T. V., Palmer, A. W. and Saini, D. P. S.: *J. Lightwave Techn.* LT-5 (1987) 7, 972.
197. Brekhovskih, L. M.: *Waves in Layered Media*. Academic Press New York, 1987.
198. Tai, H., Tanaka, H. and Yoshino, T.: *Optics Letters* 12 (1987) 6, 437.

199. Jones, B. E. and Philp, G. S.: *Proc. Europ. Conf. Sensors and Applicat. 1983*, London, Inst. of Physics, p. 86.
200. Thornton, K. E. B., Uttamchandani, D. and Culshaw, B.: *Electronics Letters* 24 (1988) 10, 573.
201. Andres, M.V. et al.: *Electronics Letters* 23 (1987) 15, 774.
202. Varnham, M. P. et al.: *J. Lightwave Techn.* LT-1 (1983) 332.
203. Dyott, R. B. et al.: *Electron. Letters* 15 (1979), 380.
204. Göpel, W.: *Techn. Messen* 52 (1985) 2, 47; 3, 92; 5, 175.
205. Göpel, W. and Oehme, F.: *hard and soft. Fachbeilage Mikroperipherik*, I–III, March 1987.
206. Göpel, W.: *Sensors and Actuators* 16 (1989), 167.
207. Nylander, C.: *J. Phys. E: Sci. Instr.* 18 (1985), 736.
208. Hulley, B.: *Measurement and Control* 21 (1988), 44.
209. Mosley, P. T. and Tofield, B. C.: *Solid State Gas Sensors*, Adam Hilger, Bristol, 1987.
210. Wing-Fong, C. and Zucholl, K.: *Technische Rundschau* 42 (1988), 154.
211. *Proc. 14th Automotive Materials Conference 1986*. Am. Cer. Soc. Inc. 1987.
212. Janata, J. and Huber, R.J.: *Solid State Chemical Sensors*, Academic Press, New York 1985.
213. Edmonds, T. E.: *Chemical Sensors*, Blackie, Glasgow, 1988.
214. Göpel, W.: *Progr. in Surf. Sci.* 20 (1985) 1, 9.
215. Heiland, G.: *Sensors and Actuators* 2 (1982), 343.
216. Morrison, S. R.: *Sensors and Actuators* 2 (1982), 329.
217. Heiland, G.: *VDI-Berichte* 509 (1984), 223.
218. Heiland, G. and Kohl, P.: *Sensors and Actuators* 8 (1985), 223.
219. Clifford, P. K.: *Proc. Int. Meeting on Chemical Sensors 1983*, Fukuoka, p. 135.
220. McAleer, J. F. et al.: *J. Chem. Far. Trans.* 1 (1987), 1323.
221. McAleer, J. F. et al.: *AERE-R 13055*, Materials Dev. Div. Harwell (UK), March 1988.
222. Williams, D. E., Stoneham, A. M. and Moseley, P. T.: *Proc. EUROSENSOR II 1988*, Enschede, Netherlands, p. 150.
223. Kohl, D. and Heiland, G.: *Proc. EUROSENSOR I 1987*, Cambridge (UK), p. 137.
224. Hübner, H. P. and Obermeier, E.: *Techn. Messen* 52 (1985) 2, 137.
225. Watson, J.: *Sensors and Actuators* 5 (1984), 29.
226. Logothetis, E. M. and Kaiser, W. J.: *Sensors and Actuators* 4 (1983), 33.
227. Katayma, K., Akiba, T. and Yanagida, H.: *Proc. Int. Meeting on Chem. Sensors 1983*, Fukuoka, p. 433.
228. Nitta, T., Terada, Z. and Hayakawa, S. J.: *Amer. Cer. Soc.* P3 (1980), 259.
229. Shastri, A. G., Schwank, J.: *Appl. Surface Sc.* 29 (1987), 341.
230. Vartanyan, A. T.: *Acta Physiochemija* (USSR) 22 (1947), 201.
231. Gutman, F. and Lyons, L. E.: *Organic Semiconductors* Wiley, New York 1967.
232. Hamann, C.: *messen steuern regeln* 31 (1988) 8, 345.
233. Bott, B. and Jones, T. A.: *Sensors and Actuators* 5 (1984), 43.
234. Wilson, A. and Collins, R. A.: *Sensors and Actuators* 12 (1987), 389.
235. Ogura, K. and Yamasaki, S.: *J. Appl. Elektrochem.* 15 (1985), 279.
236. Nylander, C. et al.: *Proc. Int. Meeting on Chem. Sensors 1983*, Fukuoka, p. 203.
237. Wiemhöfer, H. D. et al.: *Proc. SENSOR '88*, Part A, Nuremberg, p. 97.
238. Yamamoto, N., Tonomura, S. and Matsuoka, T.: *Surface Sc.* 92 (1980), 400.
239. Lundström, I. and Söderberg, S.: *Sensors and Actuators* 2 (1981/82), 105.
240. Lundström, I. et al.: *J. Appl. Phys.* 46 (1975), 3876.

241. Kaneyasu, M. *et al.*: *IEEE Trans. on Compon., Hybrids, and Manufactur. Tech.* CHMT-10 (1987) 2, 267.
242. Dobos, K., Mokwa, W. and Zimmer, G.: *hard and soft. Fachbeilage Mikroperipherik*, XIV–XVI, March 1987.
243. Janata, J. and Josowicz, M.: *Anal. Chem.* 58 (1986), 514.
244. Böhm, H. and Fleischmann, R.: *Sensoren 86/87*, VDI-Verlag, p. 50.
245. Profos, P.: *Ind. Messtechnik*, Vulcan Verlag, Essen 1974.
246. Buch, R.: *Sensors and Actuators* 1 (1981), 197.
247. Moody, G. J. and Thomas, J. D. R.: *Ion-selective Electrodes in Analytical Chemistry*. Vol. 1, Plenum Press, New York 1978.
248. Engels, J. M. L. and Kuypers, M. H.: *J. Phys. E: Sc. Instr.* 16 (1983), 987.
249. Clark, L. C. *et al.*: *Appl. Physiol.* 6 (1953), 189.
250. Serak, L.: *The Science of the Total Environment* 37 (1984), 107.
251. Jones, A., Moseley, P. and Tofield, B.: *Chem. in Britain*, August 1987, 749.
252. Ullmann, H.: *msr* 31 (1988) 8, 356.
253. Möbius, H. H.: *Z. phys. Chem.* 230 (1965), 396.
254. Fischer, W. and Rohr, F.: *Chem. Ing. Tech.* 50 (1978), 303.
255. Logothetis, F. M.: *Ceram. Eng. Sci. Proc.* 9–10 (1987), 1058.
256. Sasayama, T. *et al.*: *Ceram. Eng. Sci. Proc.* 9–10 (1987) 1074.
257. Wiedemann, H. M.: *VDI-Berichte* 578 (1986), 129.
258. Williams, D. E. and Moseley, P. T.: *Measurement and Control* 21, March (1988), 48.
259. Saji, K. *et al.*: *J. Electrochem. Soc.* 135 (1988) 7, 1686.
260. Velasco, G.: *Proc. 2. Internat. Meeting on Chem. Sensors* Bordeaux (1986), p. 79.
261. Shimizu, Y. *et al.*: *Sensors and Actuators* 14 (1988), 319.
262. Möbius, H. M.: *Z. f. Chemie* 9 (1969) 4, 158.
263. Teske, K. and Gläser, W.: *Microchimica Acta* 1 (1975), 653.
264. *Applikation FE-Messgeräte systemursalyt G*, VEB Junkalor Dessau.
265. Jahnke, D. *et al.*: *Elektrochemische Sauerstoffmessung in der Metallurgie*, Verlag Stahleisen, Düsseldorf 1985.
266. Takeuchi, T.: *Sensors and Actuators* 14 (1988), 109.
267. Maclay, C. J., Buttner, W. J. and Stetter, J. R.: *IEEE Trans. on Electr. Dev.* ED-35 (1988) 6, 793.
268. Bergveld, P.: *IEEE Trans. Biomed. Eng.* BME-17 (1970), 70.
269. Saamann, A. A. and Bergveld, P.: *Sensors and Actuators* 7 (1985), 75.
270. Klein, M.: *Hard and Soft. Fachbeilage Mikroperipherik*, XII–XIII, March 1987.
271. Smith, R. L. and Collins, S. D.: *IEEE Trans. Electr. Dev.* ED-35 (1988) 6, 787.
272. Pham, M. T. *et al.*: *msr* 31 (1988) 8, 365.
273. Bousse, L., deRooij, N. F. and Bergveld, P.: *IEEE Trans. Electr. Dev.* ED-30 (1983), 1263.
274. Voorthuyzen, J. A., van Vossen, H. and Bergveld, P.: *Proc. EUROSENSOR I* 1987, Cambridge (UK), p. 43.
275. Sallardene, J. *et al.*: *Abstracts TRANSDUCER S '89*, Montreux, p. 66.
276. Bergveld, P.: *Sensors and Actuators* 17 (1989), 3.
277. Igarashi, I. *et al.*: *Abstracts TRANSDUCER S '89*, Montreux, p. 3.
278. Pham, M. T. *et al.*: Research report ZfK Rossendorf KFE 5/88 (1988).
279. Sibbald, A.: *Sensors and Actuators* 7 (1985), 23.
280. Wolfbeis, O. S.: *Ö. Chem. Z.* 12 (1988), 350.
281. Narayanaswamy, R. and Sevilla, F.: *J. Phys. E: Sc. Instr.* 21 (1988), 10.

282. Tenge, B. *et al.*: *InTech* October (1987), 29.
283. Smith, A.M.: *Electronics and Power* November/December (1986), 811.
284. Smith, A. M.: *Analytical Proc.* 24 (1987) 1, 15.
285. Scheggi, A. M.: *J. Opt. Sensors* 2 (1987) 1, 3.
286. Martin, M. J. *et al.*: *Medical and Biological Eng. and Computing* November (1987), 597.
287. Sharma, A. and Wolfbeis, O. S.: *Appl. Spectroscopy* 42 (1988) 6, 1009.
288. Haase, T.: *US Patent* 4201222 (1980).
289. Lowe, C. R.: *Trends in Biotechnology* 2 (1984) 3, 59.
290. Turner, A. P. F., Karube, I. and Wilson, G. S.: *Biosensors: Fundamentals and Applications*, Oxford University Press, Oxford 1987.
291. Scheller, F. *et al.*: *Biosensors*, Barking 1 (1985) 2, 135.
292. Schubert, F. and Scheller, F.: *Tag. Ber. Akademie Landwirtschaftswiss. DDR, Berlin* 265 (1988), p. 179.
293. Vadgama, P.: *Abstracts TRANSDUCER '89*, Montreux, p. 1.
294. Turner, A. P. F.: *Sensors and Actuators* 17 (1989), 433.
295. Clark, L. C. and Lyons, S. C.: *Ann. N.Y. Acad. Sci.* 102 (1962), 29.
296. Göpel, W.: *Techn. Mitteilungen* 80 (1987) 8, 529.
297. Scheller, F.: *Hard and Soft*, October 1987, VI.
298. Baker, M. and Vadgama, P.: *Measurement and Control* 21 (1988), 53.
299. Watson, L. D. *et al.*: *Biosensors 3* (1987), 101.
300. Danielsson, B. *et al.*: *Sensors and Actuators* 13 (1988) 2, 139.
301. Muramatsu, J. *et al.*: *Anal. Chim. Acta* 188 (1986), 257.
302. Nakamoto, S. *et al.*: *Sensors and Actuators* 13 (1988) 2, 165.
303. Seiyama, T., Yamozoe, N. and Arai, H.: *Sensors and Actuators* 4 (1983), 85.
304. Nitta, T. and Hayakawa, S.: *IEEE Trans. Comp., Hybrids Man. Tech.* CHMT-3 (1980), 237.
305. Hooper, A. *et al.*: AERE-R 11208 Harwell (UK), March 1984.
306. Sadaoka, Y. *et al.*: *J. Materials Sc.* 22 (1987), 2975.
307. Sakai, Y. *et al.*: *Sensors and Actuators* 13 (1988), 243.
308. Sakai, Y. *et al.*: *Sensors and Actuators* 16 (1989), 243.
309. Voigt, J., Krusche, J. and Fiedler, K.: *Proc. 'Fortschritte Del Sensortechnik'*, Part 2, Jena (1989), p. 21.
310. Hamann, C. and Kampfrath, G.: *Vacuum* 34 (1984) 12, 1053.
311. Spichinger, R.: *Proc. SENSOR 88*, Part B, Nuremberg, p. 87.
312. Eder, F. X.: *Moderne Messmethoden der Physik*, Deutscher Verlag der Wissens. Berlin (1952).
313. Hart, K.: *Einführung in die Messtechnik*, Verlag Technik, Berlin 1989.
314. Hirzinger, G. and Dietrich, J.: *Techn. Messen* 53 (1986) 7/8, 287.
315. Krimmel, W.: *messen prüfen automatisieren* November (1985), 614.
316. Kautsch, R.: *Elektropraktiker, Berlin* 38 (1984) 9, 308.
317. Hellwig, R.: *messen prüfen automatisieren* May (1986), 273, June (1986), 340.
318. Neske, E. and Dumbs, A.: *Proc. SENSOR '85*, Karlsruhe.
319. Sorge, G. and Hauptmann, P.: *Ultraschall in Wissenschaft und Technik*. Teubner Verlag, Leipzig 1984.
320. Bergmann, W.: *Der Ultraschall*, St. Hirzel Verlag, Stuttgart 1954.
321. Suttle, N. A.: *GEC J. Res.* 5 (1987) 3, 141.
322. Manthey, W. and Kroemer, N.: *Proc. '5. Fachtagung Anwend. Mikrorechner in der Mess- und Automat. technik' 1989*. Magdeburg, p. 207.

323. Leibson, S. H.: *EDN Nov.* 28 (1985), 75.
324. Walker, H.: *Proc. SENSOR '88*, Part B, Nuremberg, 1988, p. 391.
325. Führer, V. and Eggers, C.: *messen prüfen automatisieren* October (1986), 620.
326. Biehl, K. E. and Müller-Gronau, W.: *Techn. Messen* 55 (1988) 10, 367.
327. Hauck, A.: *VDI-Berichte* 677, VDI-Verlag, Düsseldorf 1988, p. 41.
328. Hauptmann, P.: *VDI-Berichte* 677, VDI-Verlag, Düsseldorf 1988, p. 35.
329. Hauptmann, P.: *Int. J. Adv. Manufacturing Techn.* 2 (1987) 3, 40.
330. Dobler, K. W. and Hachtel, H.: *Proc. SENSOR '85*, Karlsruhe.
331. Göpel, W.: *Technol. für die chem. und biochem. Sensorik*. VDI/VDE Technol. zentrum Informationstechnik GmbH, Berlin (1988).
332. Decroux, M.: *Abstracts TRANSDUCER '89*, Montreux, p. 2.
333. Herb, J.A. et al.: *Abstracts TRANSDUCER '89*, Montreux p. 274.
334. Knutti, J. W.: *Proc. ASICT*, Toyohashi (Japan) 1989.
335. Kamimura, K., Kimura, N. and Onuma, Y.: *Abstracts TRANSDUCER '89*, Montreux, p. 270.
336. Gutierrez Monreal, F. J. and Mari, C. M.: *Sensors and Actuators* 12 (1987), 129.
337. Miasik, J. J. et al.: AERE-R 12027, Harwell (UK), December 1986.
338. Kawai, H.: *Jpn. J. Appl. Phys.* 8 (1969), 975.
339. Halvorsen, D. L.: *Proc. SENSOR 88*, Part A, Nuremberg, p. 29.
340. Mequio, C. et al.: *Sensors and Actuators* 14 (1988), 1.
341. Gerliczy, G. and Betz, R.: *Sensors and Actuators* 12 (1983), 207.
342. Danz, R.: *Wiss. u. Fortschritt* 36 (1986) 6, 139.
343. Patterson, R. W. and Nevill, G. E.: *Robotica* 4 (1986), 27.
344. Dario, P. and DeRossi, D.: *IEEE Spectrum Aug.* (1985), 86.
345. Luborsky, F. E.: *Amorphous Metallic Alloys*, Butterworths, London, 1983.
346. Boll, R. and Hinz, G.: *Techn. Messen* 52 (1985) 5, 189.
347. Förster, F.: *Z. Metallkunde* 46 (1955), 358.
348. Forkel, W.: *Sensoren 86/87*, VDI Verlag, Düsseldorf 1986.
349. Balzerowski, C. and Rost, R.: *Proc. 'Forts. d. Sensortechnik'*, *Jena*, 1989, p. 76.
350. Dahle, O.: *ASEA-Zeitschrift* 5 (1960), 155.
351. Mori, K. and Korekoda, S.: *IEEE Trans. Magnetics* MAG-14 (1978), 1071.
352. Harada, K. et al.: *IEEE Trans. Magnetics* MAG-18 (1982), 1767.
353. Sasada, I. et al.: *IEEE Trans. Magnetics* MAG-20 (1984), 951.
354. Juckenack, D. and Molnar, J.: *Techn. Messen* 53 (1986) 5, 242.
355. Török, E. and Hansch, G.: *J. Magn. Magn. Mat.* 10 (1979), 303.
356. Shirae, K. and Honda, A.: *IEEE Trans. Magnetics* MAG-17 (1981), 3151.
357. Wiegand, J. R.: *Method of Manufacturing Bistable Magnetic Devices*, US Patent 3892118 (1975).
358. Gevatter, H. J. and Kuers, G.: *Technisches Messen* 51 (1984) 4, 123
359. Mohri, K., Takeuchi, S. and Fujimoto, T.: *IEEE Trans. Magnetics* MAG-17 (1981), 3370.
360. Turner, J. D.: *IEEE Proceedings* 135, Part. D, (1988) 5, 334.
361. Middelhoek, S. et al.: *Proc. TRANSDUCER '87*, Tokyo, p. 17.
362. Sugiyama, S. et al.: *Sensors and Actuators* 4 (1983), 113.
363. Jordan, G. R.: *J. Phys. E: Sc. Instr.* 18 (1985), 729.
364. Langdon, R. M.: *J. Phys. E: Sc. Instr.* 18 (1985), 103.
365. Sauerbrey, G. Z.: *Z. Phys. Verhandl.* 155 (1959), 206.
366. Lu, C. and Czanderna, A.W.: *Appl. of Piezoelectric Quartz Crystal Microbalances*, Elsevier, Amsterdam, 1984.

367. Miller, J. G. and Bolef, D. I.: *J. Appl. Phys.* 39 (1968), 4589.
368. Lu, C. and Lewis, O.: *J. Appl. Phys.* 43 (1972), 4385.
369. Benes, F., Schmid, M. and Thorn, G.: *Proc. SENSOR '88*, Nuremberg, p. 349.
370. King, W. H.: *Proc. 25th Annual Symp. on Frequency Control* April (1971), p. 104.
371. Bucur, R. V.: *Rev. Roum. Phys.* 19 (1974) 8, 779.
372. Hlavay, J. and Guilbault, G. G.: *Anal. Chem.* 50 (1978) 7, 980.
373. Sazkum, L. Z. et al.: *Izmer. Tekh.* (USSR) 25 (1982) 10, 52.
374. Guilbault, G. G., Kristoff, J. and Owen, D.: *Anal. Chem.* 57 (1985), 1754.
375. Berg, P. et al.: *Proc. TRANSDUCER '89*, Montreux, p. 308.
376. Ziegler, H.: *Techn. Messen* 54 (1987) 4, 124.
377. Schaudel, H. et al.: *atp* 30 (1988) 5, 219.
378. D'Amico, A. and Verona, E.: *Sensors and Actuators* 17 (1989), 55.
379. Wohltjen, H.: *Proc. TRANSDUCER '87*, Tokyo, 471.
380. Thomä, R. and Kabitzsch, K.: *rfe* 34 (1985) 8, 480.
381. Huang, P. H.: *IEEE Trans. on Electron Dev.* 35 (1988) 6, p. 744.
382. Handen, D. et al.: *Proc. IEEE Ultrasonics Symp.* (1981), p. 148.
383. Cullen, D. E. and Montress, G. K.: *Proc. IEEE Ultrasonics Symp.* (1980), p. 696.
384. Dias, J. F.: *Hewlett Packard J.* 32 (1981) 12, 18.
385. Bonbrake, T. B. et al.: *Proc. IEEE Ultrasonics Symp.* (1985), p. 591.
386. Joshi, S. G.: *Rev. Sc. Instr.* 54 (1983) 8, 1012.
387. Ishido, M. et al.: *IEEE Trans. Instr. Meas.* IM-36 (1987) 1, 83.
388. Ahmad, N.: *Proc. IEEE Ultrasonics Symp.* (1985), p. 483.
389. Kovnovich, S. and Harnan, E.: *Rev. Sci. Instr.* 48 (1977) 7, 920.
390. Wohltjen, H. and Dessy, R. E.: *Anal. Chem.* 51 (1979) 9, 1458.
391. Wenzel, S. W. and White, R. M.: *IEEE Trans. Electron. Dev.* 35 (1988) 6, 735.
392. Wenzel, S. W. and White, R. M.: *Abstr. TRANSDUCER '89*, Montreux, p. 214.
393. Amberger, E. and Vögtlin, B.: *atp* 30 (1988) 7, 224.
394. Fernisse, E. P. et al.: *IEEE Trans. Ultrasonics, Ferroelectrics Frequ. Control* 35 (1988) 3, 323.
395. *Electronics*, June (1986), 28.

Index

acceleration sensor, 61, 63–5
amorphous metals, 174–82
amperometric sensors, 127, 137, 147
analyte, 139, 143, 149
angle-of-rotation sensor, 62
array, 6

barber's pole configuration, 73–4
BCCD, 45
bioelectrode, 147–8
biological sensor, 115, 128, 143–9, 195
black body radiator, 88–9
black layer, 21
bolometer, 21, 168

capacitive sensor, 30, 60, 151, 157–9
carrier domain magnetic field sensor, 58–60
carrier mobility, 28
CCD array, 44
CCD element, 44
CHEMFET, 119, 138
  suspended gate, 125, 141
chemical sensors, 77, 78, 115–53, 195
Clark cell, 130–1
collector current, 19, 39
colour sensor, 40–2
condensation point, 128, 150, 195
condensation point sensor, 153
conduction band, 35
conductivity sensor, 119, 120–4
Coriolis flowmeter, 199
cross-sensitivity, 93, 116, 117, 120–1, 125
CVD, 69

density sensor, 197–8
dielectric field sensor, 157
digital sensor, 28, 62, 185

distance sensor, 84
distributed anti-Stokes Raman thermometry, 107–8

eddy current sensor, 167
electrochemical sensor, 119, 126–33
electron mobility, 55
ENFET, 141
enzyme electrode, 146, 149
etching
  anisotropic, 12, 13, 91
  plasma, 15–16
  wet, 15–16
etch stop layer, 12, 15
extrinsic sensor, 81–2, 143

Fabry-Pérot interferometer, 95, 98
Faraday effect, 103–5
ferromagnetic material, 49, 72
FET-Hall sensor, 56
fiber optic gyroscope, 99–100
fiber optic Mach-Zehnder interferometer, 94, 96–7
fiber optic sensor
  acceleration, 91
  distance, 84
  filling level, 85
  pressure, 85, 87
  temperature, 88, 89
  vibration, 91
filling level sensor, 85–6, 163, 198
fluid refractometer, 86–7
fluorescence, 90, 143–4
  decay time, 90, 144
  sensor, 90, 143–4
forward current, 18
forward voltage, 18
frequency-analog sensor, 27–8, 185, 197

213

Index

galvanomagnetic effect, 49
galvanomagnetic sensor, 60, 61
gas chromatographer, 67
gas sensor, 111, 117, 120, 122, 128, 145, 189
  catalytic, 124
graded-index fiber, 81

Hall angle, 51
Hall constant, 50
Hall effect, 49
Hall generator, 53–6, 60
Hall plate, 53
Hall probe, 51
heterogeneous semiconductors, 47–8
humidity sensor, 77, 150–3, 187, 195
hybrid technology, 75, 202
hydrophone, 83–4, 97, 106, 173
hygrometer, 150

image sensor, 42, 44
immobilization, 148
inductive sensor, 154–7
integrated optics, 95
interface temperature sensor, 18–22
interferometer
  Fabry-Pérot, 94, 98
  Mach-Zehnder, 94, 96–7
  Michelson, 94, 98
  Sagnac, 94
intrinsic sensor, 81–2
ion-plating, 69
ion-selective electrode, 127–8, 138–9
ion-sensitive FET, 129
ISLET, 10, 138–9

k factor, 24, 30, 174

lambda probe, 117
Lamb wave sensor, 196, 197
Large-Barkhausen discontinuity, 182
limit current probe, 136
luminescence, 88

Mach-Zehnder interferometer, 94, 96–7
magnetic diode, 53, 57, 60
magnetic field sensor, 48–62, 104, 175–8
magnetoconcentration effect, 53, 56–7
magnetoelastic sensor, 178–80
magnetoresistive effect, 51, 55–6, 72, 181
magnetoresistive sensor, 72
magnetotransistor, 57–8

metal oxide sensor, 120, 121
Michelson interferometer, 94, 98
microbending pressure sensor, 86
micromechanical sensors, 63–7
micromechanical structures, 13–16
micromechanics, 10, 13
microswitch, 83
MISFET, 139
mode, 80
  dispersion, 81
monomode fiber, 81, 113
monomode sensor, 93–105
MOSFET, 42, 56
multimode fiber, 82
multimode sensor, 93–105
multiplexed system, 105
multisensor, 11, 122, 142, 196

numerical aperture, 80

optical sensors, 34
optrode, 142–5
organic semiconductor, 124
OTDR, 106–7

palladium-Mos diode, 124–5
path neutrality, 87, 93
pattern recognition, 125
pellistor, 124
phase modulation, 93–4
photoconductivity, 34
photodiode, 36
  array, 42–4
  avalanche, 38
  drift type, 40
  photoemission, 34
  position-sensitive, 42
phototransistor, 38–9
photoresist, 15
photoresistor, 35, 36
pH sensor, 86–7, 127–8
piezocapacitive effect, 22
piezoelectric effect, 159–62
piezoelectric polymer, 169–74
piezoresistive effect, 22, 23
Pirani vacuum gauge, 33
polarimetric sensor, 103
polarization modulation, 102
polarization OTDR, 109
position sensors, 42, 77, 177
potentiometric sensor, 127, 137
presence sensor, 164

PRESSFET, 31–2
pressure sensor
  capacitive, 30, 77
  fiber optic, 85, 87
  inductive, 157
  magnetoelastic, 179
  micromechanical, 63
  piezocapacitive, 23
  piezoelectric, 173
  piezoresistive, 24
  poly-Si-, 29, 70
  SAW-, 165
  smart, 27
pyroelectricity, 169
pyrometer, 16

quartz microbalance sensor, 188, 197
quartz resonance sensor, 186–92
quartz temperature sensor, 190

Raman effect, 108
Rayleigh light, 106
Rayleigh scattering, 107
Rayleigh signal, 109
Rayleigh wave, 192
receptor, 145
reference electrode, 129, 139
reference ISFET, 140
refractometer, fiber optic, 86, 87
relative humidity, 123, 150
resistance temperature sensor, 16–18
resonance sensor, 20, 65, 185–200
resonant microbridge, 66
resonant structure, 112
ring oscillator, 20, 27, 28

Sagnac effect, 99–100
Sagnac interferometer, 94
saturation core probe, 175–7
saturation current, 18
SAW sensor, 192–7
Schottky barrier, 37, 40
Schottky diode, 37–8
Seebeck effect, 21, 22
Sell transducer, 162
sensor
  distributed, 83, 105–9
  integrated, 6, 29, 66, 201

  intelligent, 6–7, 16, 75
  point, 83
sensor system, 8, 202
silicon
  amorphous, 9, 13, 22, 35, 40
  monocrystalline, 9, 18, 22, 27, 36
  poly, 9, 13, 18, 29, 69
silicon effects, 10
silicon properties, 9–11
silicon technology, 12
solid electrolyte sensor, 119, 133–8
spectral sensitivity, 42, 47, 48
spectrum-sensitive sensor, 42
speed of rotation sensor, 60, 62, 177–8,
    183–4
spreading resistance, 16–17
sputtering, 1 3, 69
statistical sensor, 28
stepped-index fiber, 80, 107
structured semiconductor sensor, 119, 124

tactile sensor, 173
temperature-frequency converter, 20
temperature sensor, 16, 70, 71, 76, 77, 88,
    106, 181, 190–2, 194
thermocouple, 16, 71
thermopiles, 21–2, 33
thick-film sensors, 75–8, 122
thin-film technology, 65–6, 68–74
three-electrode cell, 130
throughflow sensor, 63, 66, 77, 83, 195,
    199
torque sensor, 61, 157, 179–81
transverse shear oscillater, 187–8
two-electrode cell, 130

ultrasound
  absorption, 139
  sensor, 160–7
  speed, 161

valency band, 35
vibration sensor, 63
vidicon, 45

waveguide, 80
Wiegand sensor, 182–4
wire strain gauge, 24, 70